QUANTUM MECHANICS

Jasmina Novakovic

TAP

Toronto Academic Press

QUANTUM MECHANICS

Jasmina Novakovic

Toronto Academic Press
4164 Lakeshore Road
Burlington ON L7L 1A4
Canada
www.tap-books.com

Email: orders@arclereducation.com

ISBN: 978-1-77956-340-8

ABOUT THE AUTHOR

Jasmina Novaković is an accomplished physicist and educator with a rich background in applied and computer physics. Holding a bachelor's degree in Applied and Computer Physics, Jasmina has dedicated her career to making complex scientific concepts accessible and engaging. With years of experience as a physics teacher, she has a proven track record of inspiring curiosity and fostering a deep understanding of the subject among her students. Her passion for physics extends beyond the classroom. Jasmina is a committed advocate for science education and has actively participated in popularizing physics through hands-on experiments and educational initiatives. Her work includes conducting interactive sessions and workshops aimed at demystifying physics for learners of all ages. In addition to her scientific expertise, Jasmina is a creative writer and translator fluent in English and Spanish, with native proficiency in Serbian. This linguistic skill set enhances her ability to communicate complex ideas and stories across cultures. This unique combination of skills positions her as a versatile professional capable of addressing a wide range of projects with precision and cultural sensitivity. Jasmina is also committed to community service, having volunteered in education, disaster relief, and legal aid. Her diverse background in physics, writing, and community support provides a unique and enriched perspective in all her endeavors.

Table of Contents

List of Figures

PREFACE

Quantum Mechanics is a fundamental theory in physics that describes the behavior of particles at atomic and subatomic scales, where classical physics fails to explain observed phenomena. It introduces concepts such as wave-particle duality, quantization of energy, and the uncertainty principle, highlighting the probabilistic nature of particle properties and interactions. The theory employs mathematical frameworks, including wave functions and operators, to predict the outcomes of measurements and the evolution of systems over time. Quantum mechanics is crucial for understanding atomic and molecular structures, solid-state physics, and has applications in advanced technologies like quantum computing and quantum cryptography.

This book comprises chapters; the first chapter discusses the Origin of Quantum Mechanics and delves into the historical development of this field, tracing its roots from the early 20th century breakthroughs in understanding radiation and the dual nature of particles. Begins with Planck's explanation of blackbody radiation and Einstein's interpretation of the photoelectric effect. Subsequent chapters build upon this foundation, introducing the postulates of Quantum Mechanics, which define the mathematical framework for describing quantum systems and their observable properties.

Chapter three shifts the focus onto the Schrödinger Equation, which provides a mathematical basis for understanding the behavior of quantum systems in different physical contexts. While chapter four extends this understanding to Quantum Mechanics in one dimension, exploring phenomena such as barrier penetration and the implications of bound and unbound states. Angular Momentum, discussed in chapter five, introduces the formalism for describing rotational properties in quantum systems, including orbital and spin angular momentum. The Hydrogen Atom, covered comprehensively in chapter six, serves as a pivotal example for applying Quantum Mechanics to real-world scenarios, detailing energy levels, radiative transitions, and relativistic effects. Scattering Theory, explored in chapter seven, addresses the interaction of particles with potential fields, providing insights into cross-sections and scattering amplitudes.

Finally, chapter eight explores modern applications of Quantum Mechanics, focusing on Quantum Cryptography. This chapter elucidates how quantum principles are harnessed for secure communication, covering concepts such as quantum bits, quantum computing, and quantum key distribution protocols like BB84.

Throughout the book, each chapter is structured to build upon the previous one, offering readers a systematic progression from historical context to advanced applications. The aim is to equip students and researchers alike with a solid foundation in Quantum Mechanics, fostering a deeper appreciation for its theoretical underpinnings and practical implications in contemporary science and technology.

–Author

CHAPTER

1 Origin of Quantum Mechanics

LEARNING OBJECTIVES

After studying this chapter, you will be able to:

1. Understand the limitations of classical physics that prompted the creation of quantum mechanics.
2. Comprehend the significance of blackbody radiation
3. Understand the photoelectric effect and de Broglie's hypothesis
4. Comprehend the classical and quantum view of particles and waves
5. Comprehend Heisenberg's principle and understand its significance in quantum mechanics
6. Learn about Rutherford's and Bohr's model of an atom

KEY TERMS FROM THIS CHAPTER

blackbody radiation
de Broglie's hypothesis
photoelectric effect
radiation
classical physics
indeterministic
quantum mechanics

UNIT INTRODUCTION

Quantum mechanics is the fundamental theory in physics that describes the behavior of matter and energy at the smallest scales, including atoms and subatomic particles. It provides a framework for understanding phenomena that cannot be explained by classical theories, marking a significant shift from traditional mechanics. This chapter explores the key discoveries that established quantum mechanics, highlighting its impact on physics (Kaniadakis, 2002).

The chapter begins with a historical overview of experimental findings and theoretical challenges that classical physics could not address. It examines the particle nature of radiation by phenomena such as the photoelectric effect and blackbody radiation. These experiments demonstrated the need for a new approach to understanding the quantized nature of energy and the dual role of light as both a particle and a wave (Darrigol, 2009).

The de Broglie hypothesis introduced the wave nature of particles, suggesting that matter, like light, exhibits wave-like properties. This idea laid the foundation for the concept of wave-particle duality in quantum mechanics. The chapter contrasts classical and quantum views on particles and waves, highlighting the notion that particles can behave like waves and vice versa. This leads to a discussion on the Heisenberg Uncertainty Principle, which explains the inherent limitations in measuring quantum systems (Dürr et al., 1998).

Further, the chapter explores atomic transitions and spectroscopy, tracing the development from Rutherford's planetary model to Bohr's model of the hydrogen atom. These models provided essential insights into atomic structure and the discrete nature of electron energy levels. By focusing on the experiments, theories, and conceptual developments that shaped quantum mechanics, this chapter aims to provide a detailed understanding of the origins of this field (El'yashevich, 1977).

Did you know?

The term "Quantum Mechanics" was coined in the late 1920s and refers to the quantization (discrete nature) of physical properties at the atomic scale.

1.1. HISTORY OF QUANTUM MECHANICS

By the end of the 1800s, the main fields of study in physics were thermodynamics, electromagnetism, and classical mechanics. The dynamics of material bodies were predicted by classical mechanics, and radiation was studied within the appropriate framework of Maxwell's electromagnetism; radiation and matter were respectively represented in terms of waves and particles. The Lorentz force or thermodynamics provided a satisfactory explanation for how radiation and matter interacted (Ireson, 2000).

The great achievements of classical physics, including thermodynamics, classical mechanics, and the classical theory of electromagnetism, led many to conclude that the perfect explanation of nature had been discovered. It appears that the general concepts of matter and radiation can account for all observed physical phenomena. However, classical physics, which had remained virtually unchallenged, faced significant challenges at the turn of the twentieth century on two main fronts:

i. Relativistic Domain: The theory of relativity, proposed by Einstein in 1905, demonstrates the limitations of Newtonian physics when dealing with extremely high speeds, such as those approaching the speed of light (Muller, 1997).
ii. Microscopic Domain: When atomic and subatomic structures could be investigated through new experimental techniques, it became clear that many recently reported phenomena could not be satisfactorily explained by conventional physics. It became clear that classical physics is not applicable at the microscopic level, and novel concepts were needed to explain the structure of atoms and molecules, as well as the interaction of light with them (Castellani, 2019).

The boundaries of classical physics proved insufficient in understanding a range of microscopic phenomena, such as the photoelectric effect, blackbody radiation, atomic stability, and atomic spectroscopy. This led to an investigation of alternative concepts that extend beyond their limitations. In 1900, Max Planck made a groundbreaking discovery that revolutionized the field of physics (Bunge, 1956).

He successfully validated the findings of blackbody radiation experiments by proposing the idea that energy is exchanged between radiation and its environment in distinct, or quantized, increments. Based on the argument presented, it is stated that matter and an electromagnetic wave with a frequency F can only transfer energy in multiples of hF. The energy he referred to was termed a quantum, with h representing Planck's constant, a fundamental constant (Hilgevoord, 2005).

It has been demonstrated that the idea of quantizing electromagnetic radiation has important and broad implications. Planck's theory, which offered a precise explanation of blackbody radiation, sparked novel ideas and initiated a series of new discoveries that led to answers for the most important problems of the day (Hund, 1977).

In 1905, Einstein presented a comprehensive synthesis of Planck's quantum theory. In order to grasp the photoelectric effect, Einstein needed to incorporate Planck's theory of the quantization of electromagnetic waves into his understanding of light. He proposed a hypothesis, inspired by Planck's work, suggesting that light consists of tiny particles known as photons, each carrying a specific amount of energy denoted by hF, where F represents the frequency of the light (Griffiths, 1984).

Einstein's discovery of the photon concept provided an accurate and simple answer for the long-standing photoelectric problem, which was originally seen experimentally by Hertz in 1887.

Another important discovery was made possible by Niels Bohr. Niels Bohr created his model of the hydrogen atom in 1913 by combining Planck's quantum theory, Einstein's idea of photons, and Rutherford's revolutionary discovery of the atomic nucleus in 1911 (Gell & Hartle, 2012).

In this study, it was argued that atoms exist in specific energy states and that their interaction with radiation, such as the emission or absorption of radiation, only happens in discrete amounts. This phenomenon occurs because of the atom transitioning between its various energy states. This work offers a thorough explanation for various unresolved inquiries, including atomic spectroscopy and atomic stability (David, 2015).

In 1923, Compton made a significant discovery that provided compelling evidence supporting the idea that light behaves as a particle. In his research, he established the particle-like behavior of X-ray photons through the phenomenon of scattering X-rays with electrons. This behavior is characterized by the momenta hFc, where F represents the frequency of the X-rays (Wiseman, 2006).

The understanding of the particle component of waves, which involves the observation that waves exhibit particle-like behavior on a small scale, was developed through a series of important advances made by Planck, Bohr, Einstein, and Compton. These breakthroughs established the theoretical basis and undeniable experimental proof for this phenomenon. When it comes to this level of scientific inquiry, classical physics is not only unclear, but it also fails to meet the standards of excellence and comprehensive comprehension (Saunders, 1998).

In 1923, de Broglie introduced a concept that posed a challenge to classical physics. He proposed that not only does radiation behave like particles, but also material particles exhibit wave-like behavior. The experimental confirmation of this concept occurred in 1927 by Davisson and Germer, who demonstrated that material particles, such as electrons, can exhibit interference patterns, a characteristic behavior of waves (Garritz, 2013).

In the past, there existed two distinct formulations of quantum mechanics. Heisenberg devised the first formulation, known as matrix mechanics, in 1925 to explain atomic structure based on observed spectral lines. Heisenberg's hypothesis was based on the idea that energy exchange between microphysical systems can only occur in specific, discrete values known as quanta. Bohr's conception of the hydrogen atom and Planck's quantization of waves both had an impact on this concept (Bub, 2000).

He used matrices to create an eigenvalue problem that describes the energy, momentum, position, and angular momentum of microscopic systems. Once the Hamiltonian matrix is diagonalized, the system's energy spectrum and state vectors become visible. When it came to understanding the discrete light units that atoms emit and absorb, matrix mechanics proved to be quite helpful (Blum & Jähnert, 2022).

A different approach to the subject was introduced by Schrödinger in 1926. This formulation, known as wave mechanics, builds upon the de Broglie postulate. This approach, which is considered simpler compared to matrix mechanics, explains the behavior of tiny particles through a wave equation known as the Schrödinger equation. Schrödinger created a differential equation to represent the dynamics of tiny matter, in contrast to Heisenberg's matrix eigenvalue problem. The energy spectrum and wave function of the system under study can be understood by looking at the answers to this equation. Max Born proposed his probabilistic interpretation of wave mechanics in 1927. He deduced probability densities from an analysis of the wave functions' square moduli, which are Schrödinger equation solutions (Baracca et al., 2017).

1.2. RADIATIONS PARTICLE ASPECT

In classical physics, particles are described by their energy and momentum, while waves are described by their amplitude and wave vector. The wave vector $(|\;| = 2\pi/\lambda)$ indicates the direction of wave propagation. In the field of physics, particles and waves display distinct and contrasting behaviors. It is worth noting that the properties of being a "particle" and a "wave" are completely incompatible with each other. It is important to acknowledge that electromagnetic waves can transfer varying amounts of energy to particles (Koch et al., 2024).

This section will explore how the inflexible principles of classical physics ultimately proved inadequate in explaining multiple microscopic phenomena, including blackbody radiation and the photoelectric effect. Understanding these events required a shift from the rigid principles of classical physics and acceptance of a groundbreaking concept: the particle nature of radiation (Phillips et al., 1990).

1.2.1. Blackbody Radiation

The interaction between radiation and matter is an issue here. A solid item glows and releases thermal radiation when heated. The thing turns red, then yellow, then white as the temperature rises. From infrared to ultraviolet, glowing solid objects emit a constant variety of frequencies in their thermal radiation. The distribution spectrum of heated gases is continuous as opposed to the radiation released by heated gases, which has a discrete distribution spectrum made up of a few sharp (narrow), multicolored lines with no light (i.e., blackness) in between (Gallagher & Cooke, 1979).

In the latter part of the 1800s, one of the main unresolved issues was the understanding of the continuous nature of the radiation released by a blazing solid object. All attempts to use the current theories of classical physics, classical electromagnetic theory and statistical thermodynamics) to explain this phenomenon have failed miserably. The main task at hand was essentially defining the appropriate thermodynamic theory that explains the energy transfer between matter and radiation (Prokhorov et al., 2009).

When an object is exposed to radiation, a portion of it may be absorbed while the rest is reflected. A blackbody is a hypothetical object that, when illuminated from the outside, appears black because it absorbs all incident radiation. The thermal motion of the electrons on an object's surface causes it to release electromagnetic energy when it gets heated. Its temperature and frequency determine how strong the radiation is, yet the light it emits is full spectrum. When a thing is in thermal equilibrium with its environment, it releases as much energy as it takes in. Therefore, it may be concluded that a blackbody possesses the ability to absorb radiation completely and release radiation flawlessly (Cooke & Gallagher, 1980).

Remember

Planck's quantization hypothesis resolved the ultraviolet catastrophe by introducing the idea that energy is emitted in discrete packets (quanta).

A hollow cavity with a tiny hole on its surface and interior walls that precisely reflect electromagnetic radiation, such as metallic walls, can be used to create a working blackbody. After numerous reflections of the cavity's inner walls, any radiation that manages to get inside through the hole will be entirely absorbed (Smoot et al., 1977).

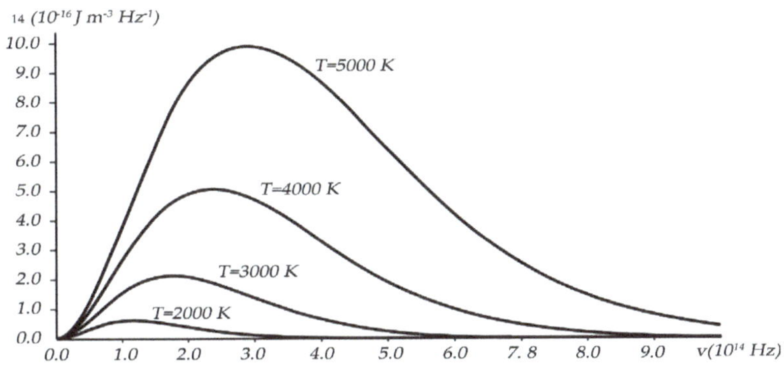

Figure 1.1. Illustration of blackbody radiations spectral energy density as function of frequency at various temperatures.

Source: Nouredine Zettili, Creative Commons License.

The cavity therefore absorbs electromagnetic radiation in a manner like that of a black body. On the other hand, blackbody radiation is the radiation that emerges from the hole when the cavity is heated to a particular temperature T. The hole serves as an excellent emitter and will soon begin to emit visible light when the temperature rises. Simply examining the spectral distribution of the radiation emitted from the aperture provides valuable insight into the radiation within the cavity. Furthermore, the concept of blackbody radiation refers to the emission of radiation from a heated hollow chamber's opening. Blackbody radiation is the term used to describe the radiation emitted by a blackbody when it is heated (Haslinger et al., 2018).

During the mid-1800s, researchers collected a significant amount of empirical data on blackbody radiation from different objects. These findings indicate that when in a state of balance, the emitted radiation possesses a clearly defined and uninterrupted distribution of energy. Each frequency is associated with an energy density that is entirely determined by the temperature of the cavity's walls and is unaffected by the object's chemical composition or shape (Figure 1.1). At a certain frequency, the energy density shows a noticeable peak that increases with temperature. Put another way, a frequency that is precisely proportional to temperature corresponds to the highest point on the radiation spectrum (Narayanaswamy et al., 2009).

This explains the color changes that an object undergoes when heated to a greater temperature, most notably from red to yellow to white. As it turned out, there was more to the blackbody spectrum than meets the eye. There were other attempts to explain the continuous character of the radiation. Two significant attempts, one in 1889 by Wilhelm Wien and another in 1900 by Rayleigh, both relied on classical physics. In 1879, J. Stefan conducted an experiment to calculate the total intensity of radiation emitted by a high-temperature object. The experiment aimed to determine the total power per unit surface area radiated by the object (Dunbar & McMahon, 1998).

$$\mathcal{P} = a\sigma T^4$$

Which is known as the Stefan–Boltzmann law, Where; σ= 5.67 × 10^{-8} W m^{-2} K^{-4} is the

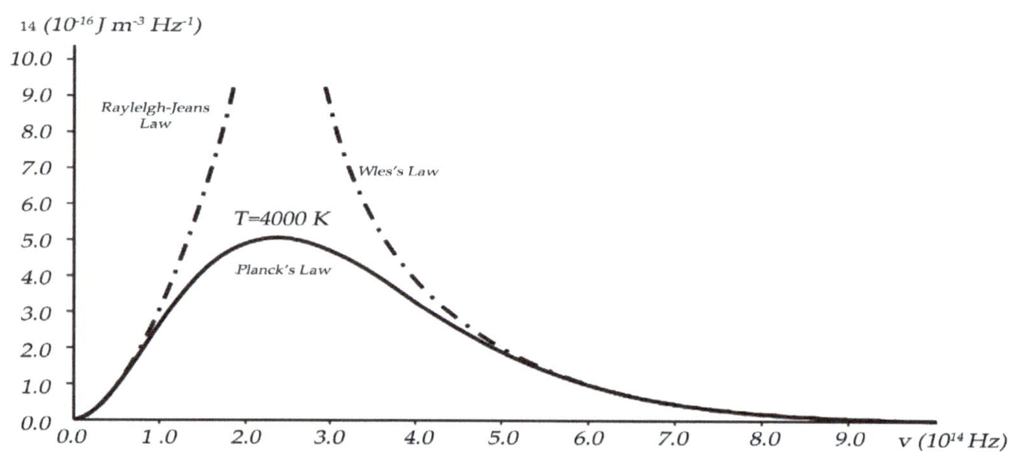

Figure 1.2. Several spectral densities comparison: The experimental and Planck distributions exhibits complete match (solid curve), but the Wien and Rayleigh-jeans distributions only partially aligns with experimental distribution.

Source: Richard Lebed, Creative Commons License.

The Stefan–Boltzmann constant, denoted by σ, is a fundamental physical constant. The coefficient a, which can range from 0 to 1, represents the emissivity of an object. For a blackbody, the coefficient is equal to 1. In 1884, Boltzmann combined the ideas of thermodynamics and Maxwell's electromagnetic theory to provide a theoretical justification for Stefan's empirical law (Maslovski et al., 2016).

1.2.1.1. Planck's Energy Density Distribution

In 1900, Planck avoided the problem of ultraviolet catastrophe and offered a precise explanation of blackbody radiation by cleverly combining Wien's rule and the Rayleigh-Jeans rule through interpolation. Planck's viewpoint diverged significantly from Rayleigh's idea that a standing wave could transfer an unlimited amount of energy to matter. Planck maintained that radiation and matter must undergo energy exchange in independent units. According to his theory, the energy of the radiation (with a frequency F) that is emitted by the oscillating charges (from the hollow walls) must be measured in discrete units that are whole number multiples of hv (Popa & Vasile, 2008).

$$E = nh\nu, \qquad n = 0,\ 1,\ 2,\ 3,\ \cdots,$$

In this context, the symbol h represents a fundamental constant. This phenomenon arises due to the correspondence between the frequency of the emitted radiation by a vibrating charged particle and the frequency of the particle's oscillation. In essence, the energy of an oscillator with a natural frequency F, which aligns with the energy of a charge vibrating at frequency F, must be a multiple of hv. It is important to note that the value of hF varies among oscillators and is influenced by the frequency of each oscillator (Dougal, 1976).

However, there are no limitations on the frequency or energy that an oscillator can have due to classical mechanics. Pendulums, mass-spring systems, and electric oscillators are examples of oscillators whose energy fluctuates constantly in relation to frequency. The formula above is referred to as Planck's postulate or the quantization rule for energy. Planck established that the average energy of an oscillator may be stated using the appropriate thermodynamic equation, providing that the energy of the oscillator is quantized:

$$\langle E \rangle = \frac{\sum_{n=0}^{\infty} nh\nu e^{-nh\nu/kT}}{\sum_{n=0}^{\infty} e^{-nh\nu/kT}} = \frac{h\nu}{e^{h\nu/kT} - 1},$$

The radiation released from a cavity's hole has an energy density per unit frequency that is determined by;

$$\boxed{u(\nu, T) = \frac{8\pi\nu^2}{c^3}\frac{h\nu}{e^{h\nu/kT} - 1}.}$$

This phenomenon is commonly referred to as Planck's distribution. It accurately matches the different experimental radiation distributions, as shown in Figure 1.2. The experimental data yields a numerical value of h as h= 6.626 × 10^{-34} Js. It is important to note that the energy density of Planck can be reformulated to determine the energy density per unit wavelength (Aatrokoski et al., 2011).

$$\boxed{\tilde{u}(\lambda, T) = \frac{8\pi hc}{\lambda^5}\frac{1}{e^{hc/\lambda kT} - 1}.}$$

In conclusion, the blackbody radiation spectrum shows how radiation is quantized, particularly how electromagnetic waves behave as particles (Verde et al., 2013). The emergence of physics of the microphysical world and the demise of classical physics were indeed predicted by the introduction of the constant h. Encouraged by Planck's successful quantization of radiation, other physicists such as Compton, Einstein, de Broglie, and Bohr skillfully modified it to account for another unsolved issue that had been unsolved for decades (Collaboration et al., 2016).

1.2.2. Photoelectric Effect

The quantization of light's energy is directly supported by the photoelectric effect. When it was shown that metals released electrons when exposed to light in 1887, Hertz made the discovery of the photoelectric effect (Figure 1.3a) (Taslidere, 2016). Additionally, prior to 1905, the following empirical principles were discovered:

i. No electron can be released, no matter how strong the radiation is, if the incident radiation's frequency is less than the metal's threshold frequency, which is established by the metal's properties. (Sorokin et al., 2007)

ii. Regardless of the radiation's low intensity, electrons will be released as soon as the frequency of incident radiation exceeds the threshold frequency v_o.

iii. The number of electrons ejected is independent of light frequency and increases with light intensity at any frequency over v_o.

iv. Rather than beam intensity, the frequency of the beam determines the kinetic energy of the released electrons. Moreover, the electron's kinetic energy increases in direct proportion to the input beam's frequency (Ossiander et al., 2018).

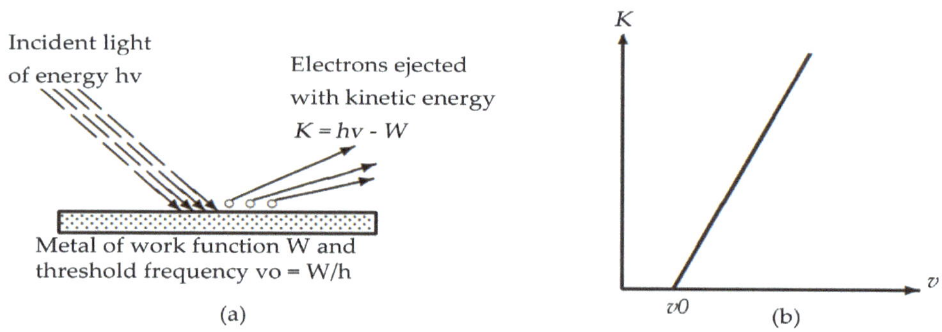

Figure 1.3. Illustration of photoelectric effect.

Note: (a) when a metal is exposed to light, electrons have the potential to be released (b) The frequency (v) of the light irradiating the metal determines the kinetic energy (K) of the electron that is leaving it.

Source: Richard Lebed, Creative Commons License.

Within the framework of a strictly classical theory of radiation, the experimental data, particularly the effect's dependence on the threshold frequency, cannot be explained. Classical physics states that matter can exchange any (continuous) amount of energy. Stated differently, since the intensity of an electromagnetic wave is proportional to the square of its amplitude, any frequency with sufficient intensity can supply the energy needed to liberate the electron from the metal (Wheaton, 1978).

However, what would occur if a weak light source was utilized? In classical physics, the electron would continue to absorb energy steadily until it reached a certain threshold, at which point it would exit the metal. For this argument to be valid, it suggests that when utilizing extremely weak radiation, the occurrence of the photoelectric effect would be significantly delayed, potentially lasting for hours, until an electron gradually built up the required energy. This conclusion, on the other hand, contradicts experimental observation completely (De Ninno et al., 2020).

Even though the light source used in the experiments was so weak that it would have taken hours for an electron to gather the energy necessary for its ejection, some electrons were seen to leave the metal right away. Subsequent tests revealed that electrons cannot be removed from the metal by intensity (brightness) improvement alone. However, electron emission begins instantaneously even at very low intensities when the incident radiation frequency is increased beyond a threshold. These experimental findings demonstrate how incorrect the classical physics-based theory of the electron's continual absorption of energy or its steady accumulation of energy (Hulme, 1931).

The electron will only absorb energy in discrete units of energy hv, regardless of the intensity of the incident radiation. If the energy of a photon (hv) is greater than the work function (W) of a metal, which is the minimum energy required to remove an electron from the metal, the electron will be ejected from the metal. Therefore, it is impossible for an electron to be emitted from the surface of the metal unless the energy of the incident light (hv) (Fano et al., 1959) is greater than the work function (W).

$$\boxed{h\nu = W + K,}$$

where K is the electron's kinetic energy as it leaves the substance. As shown in Figure 1.3b, the above equation, which was obtained by Einstein, provides the correct explanation for the experimental finding that the ejected electron's kinetic energy grows linearly with the incidence frequency F:

$$K = h\nu - W = h(\nu - \nu_0),$$

Where; v_0= W/h is referred to as the metal's cutoff or threshold frequency. Furthermore, this relationship makes it abundantly evident why no electron can leave the metal until $v > v_0$: regardless of radiation intensity, the photoelectric effect cannot happen when $v < v_0$ because kinetic energy cannot be negative. The excess energy $h(v - v_0)$ from the incident radiation provides the expelled electrons with their kinetic energy (Spagnolo et al., 2020).

One way to determine the kinetic energy of the emitted electrons is through experimental methods. The experimental arrangement, developed by Lenard, involves a photoelectric metal (cathode) positioned alongside an anode within a glass tube that has been emptied of air. When light interacts with the cathode's surface, the resulting electrons are drawn towards the anode, creating a photoelectric current. It has been noted that the photoelectric current produced directly correlates with the incident radiation intensity. Nevertheless, the frequency of the radiation affects the electrons' speed more than its intensity (Christensen, 1976).

One easy way to measure the kinetic energy of electrons is to reverse the terminals of a voltage source that has adjustable settings. The released electrons cannot get to the anode when the voltage difference V across the tube is reversed. The only electrons that may effectively reach the negative plate and contribute to the current are those whose kinetic energy is larger than the absolute magnitude of the potential difference, $e|V|$ (Kovačević & Djordjevich, 2006).

By adjusting the voltage, we can reach a specific value known as the stopping potential (Vs). At this point, all electrons, including the most energetic ones, are redirected before they can reach the collector (Güémez & Fiolhais, 2018). As a result, the flow of photoelectric current comes to a complete halt. The relationship between the stopping potential Vs and the electrons' kinetic energy can be expressed as $e|V_s|=1/2m_ev^2$. Therefore, the equation mentioned can be rewritten as eVs= hv W or

$$V_s = \frac{h}{e}\nu - \frac{W}{e} = \frac{hc}{e\lambda} - \frac{W}{e}.$$

The plot of Vs against frequency has a linear shape, like Figure 1.3b, with the slope now determined by h/e. The relationship between the stopping potential and the frequency of the incident radiation is clearly proportional. In 1916, Millikan conducted a methodical experimental validation of Einstein's photoelectric theory (Wong et al., 2011).

He generated a comprehensive compilation of photoelectric data through using several types of metals. He confirmed that the equation provided by Einstein accurately matched his facts. Furthermore, Millikan discovered that the empirical value he derived for h, by measuring the slope h/¬e of the equation shown above (Figure 1.3b) (Compton & Richardson, 1913), is approximately equal to Planck's constant with an experimental error of 0.5%.

To summarize, the photoelectric effect offers convincing evidence for the particle-like nature of electromagnetic radiation (Schirripa et al., 2019).

1.3. WAVE ASPECT OF PARTICLES

1.3.1. De Broglie's Hypothesis

As mentioned earlier, in the photoelectric effect, radiation demonstrates both particle-like characteristics and its wave nature. In 1923, de Broglie expanded upon the concept of wave-particle duality by proposing that this phenomenon is not limited to radiation alone, but rather applies universally to all material particles. This means that all particles should exhibit both wave-like and particle-like characteristics. It is essential to note that the phenomenon of wave-particle duality is not limited to light but also applies to matter (Frieden & Soffer, 2009).

The momentum of a photon, which can be expressed as $p=hv/c = h/\lambda$. This relation can be extended to any material particle that has a nonzero rest mass: Every material particle with momentum exhibits wave-like behavior, known as matter waves. The wavelength λ and wave vector of these waves are determined by the particle's speed and mass (Catillon et al., 2008).

$$\lambda = \frac{h}{p}, \quad \vec{k} = \frac{\vec{p}}{\hbar}$$

in which $h=h/2\pi$. The de Broglie relation, as stated above, establishes a connection between a particle's momentum and the wavelength and wave vector of the wave that corresponds to it.

1.3.2. Experimental Configuration of De Broglie's Hypothesis

In 1927, Davisson and Germer conducted experiments to confirm de Broglie's theory. Later, Thomson produced electron interference patterns (Brida et al., 2002).

1.3.2.1. Davisson-Germer Experiment

A 54-electron volt (eV) electron beam was focused on a nickel (Ni) crystal by Davisson and Germer in an experiment that resulted in the electrons scattering. The electron source and detector were positioned symmetrically about the crystal's normal, as shown in Figure 1.4. This set-up is comparable to the Bragg configuration that is used in grating-based X-ray diffraction. Although the electrons are dispersed from the crystal in all directions, Davisson and Germer found that the intensity peaked at an angle of Θ =35° and that the electrons mostly scatter in certain directions (Field, 2013), with a maximum scattering angle of Θ =50°.

Did you know?

Louis de Broglie's idea that particles can exhibit wave-like properties was initially met with skepticism. His doctoral thesis, however, was so compelling that it earned him a Nobel Prize in Physics in 1929.

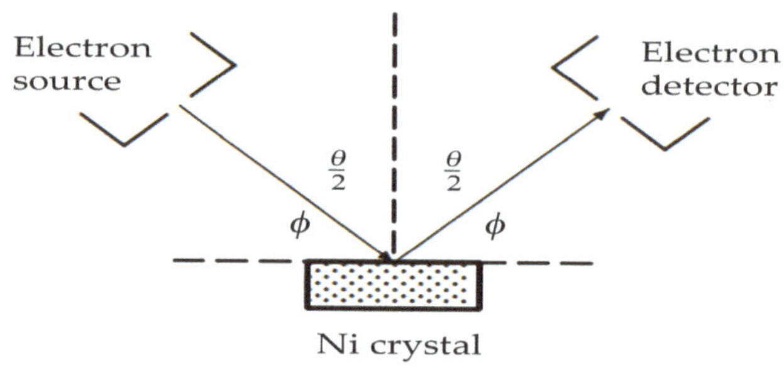

Figure 1.4. Schematic of Davisson-Germer experiment.

Source: Jearl Walker, Creative Commons License.

They demonstrated that the pattern remained consistent even when the beam's intensity was sufficiently low to send the incident electrons individually. Constructive interference of the scattered electrons is the only possible outcome. The reflected electrons created diffraction patterns identical to Bragg's X-ray diffraction by a grating, rather than the scattered distribution pattern caused by material particles (Schmidt, 2019). The Davisson-Germer experiment's scattered electron peak intensity corresponds to the Bragg formula's initial maximum (n=1).

$$n\lambda = 2d \sin \phi,$$

The angle between the reflecting planes and incident ray of the crystal is represented by the variable M, while the distance between the Bragg planes is represented by the variable d. Conversely, Θ stands for the angle that separates the scattered beams from the incident. The equation d = D sin Θ can be utilized to calculate the spacing d, which represents the distance D between consecutive atomic layers in the crystal (McKagan et al., 2008).

The value of d for a nickel crystal is 0.091 nm, given that D is equal to 0.215 nm. Given that there is only one peak observed at Θ=50° for a beam of electrons with a kinetic energy of 54 eV, and using the equation 2Φ + Θ = π and the relationship sinΦ= cos (Θ /2) (Figure 1.4), we may calculate the wavelength of the scattered electrons (Weiss, 2022).

$$\lambda = \frac{2d}{n}\sin\phi = \frac{2d}{n}\cos\frac{1}{2}\theta = \frac{2 \times 0.091\ \text{nm}}{1}\cos 25^\circ = 0.165\ \text{nm}.$$

Let us now calculate the value of λ that results from the de Broglie link. One can show that de Broglie wavelength, since the electrons' kinetic energy is K=54 eV and their momentum $p = \sqrt{2m_e K}$ is proportional to their rest mass energy ($m_e c^2$= 0.511 MeV) and ħc = 197.33 eV nm:

$$\lambda = \frac{h}{p} = \frac{h}{\sqrt{2m_e K}} = \frac{2\pi \hbar c}{\sqrt{2m_e c^2 K}} = 0.167\ \text{nm},$$

It has been observed that the electrons that were dispersed in Davisson-Germer experiment created interference patterns which were exactly the same as those produced by Bragg's X-ray diffraction. The Bragg formula accurately predicts the interference fringes of electrons. Therefore, electron motion with momentum must be described using a plane wave (Halim et al., 2020).

$$\psi(\vec{r}, t) = Ae^{i(\vec{k}\cdot\vec{r} - \omega t)} = Ae^{i(\vec{p}\cdot\vec{r} - Et)/\hbar}$$

If A is a constant, is the plane wave's wave vector, and ω is its angular frequency. De Broglie's relations are used to connect the wave's parameters, and ω, to the momentum of the electron and energy E: =/h.

1.3.2.2. Thomson Experiment

In an experiment conducted by Thomson, a polycrystalline thin film was used to diffract electrons (Figure 1.5). Additionally, diffraction fringes were noted. This outcome validated electron wave behavior again. The Davisson–Germer experiment has motivated others to produce diffraction patterns with a variety of particles (Yamada et al., 2010). Larger and larger particles, including protons, neutrons, helium atoms, and hydrogen molecules, produced interference patterns. Recent diffraction investigations of carbon 60 (C60) molecules' de Broglie wave interference at a material absorption grating provided evidence in favor of the theory that each C60 molecule only interferes with itself (a C60 molecule is nearly a classical object) (Thomson, 1928).

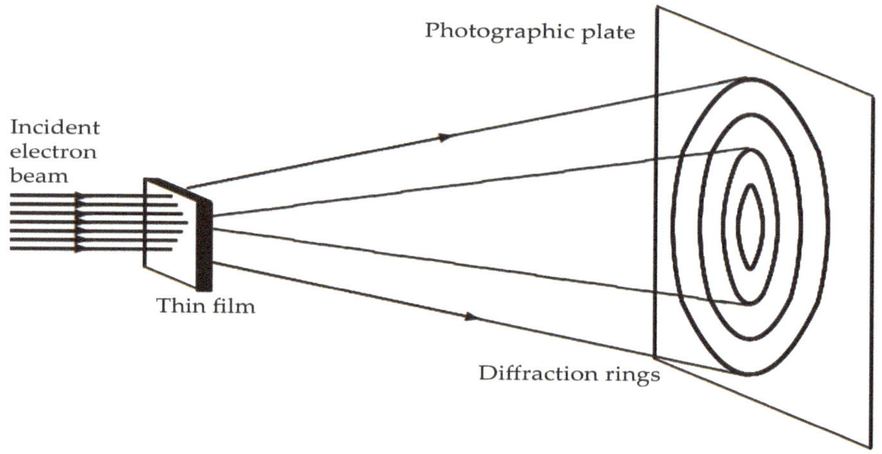

Figure 1.5. Schematic of Thomson experiment.

Source: Jearl Walker, Creative Commons License.

1.4. PARTICLES VERSUS WAVES

The properties of waves and particles in context of quantum and classical physics will be discussed in this section. These phenomena are examined using the double-slit experiment. It consists of a wall with two slits, a source (which could be a source of material particles or waves), and a back screen with counters to record what passes through the slits (DiBenedetto et al., 2018).

1.4.1. Classical View of Waves and Particles

Particles and waves are mutually exclusive in classical physics; they behave entirely differently. The position vector r (t) is all that is needed to fully characterize a particle, whereas the amplitude and phase are needed to fully characterize a wave. For example, wave functions φ = (,t) can be used to represent three-dimensional plane waves (Eliezer, 1948).

$$\psi\left(\vec{r},t\right)=Ae^{i\left(\vec{k}.\vec{r}-\omega t\right)}=Ae^{i\phi}$$

where Φ represents the wave's phase and A stands for the wave's amplitude (is the wave vector, and ω is the angular frequency). Lets recall definition of φ in the context of physics, specifically in relation to wave intensity, which can be expressed as is $I= |\varphi|^{2}$.

Examine three distinct experiments shown in Figure 1.6, where a source S releases a continuous flow of bullets. It is believed that the bullets are unbreakable and hence reach the screen in similar masses. In the initial trial, only slit S1 is unobstructed. The intensity recorded on the screen, denoted as $I_1(y)$, represents the number of projectiles that reach a specific spot y every second. Let's consider the intensity, denoted as $I_2(y)$, that is measured on the screen during the second experiment when only S2 is open. In the third experiment, the total intensity measured on the screen behind the two slits should be equal to the sum of I_1 and I_2, assuming that both S1 and S2 are open (Wetterich, 2010).

$$I\left(y\right)=I_1\left(y\right)+I_2\left(y\right).$$

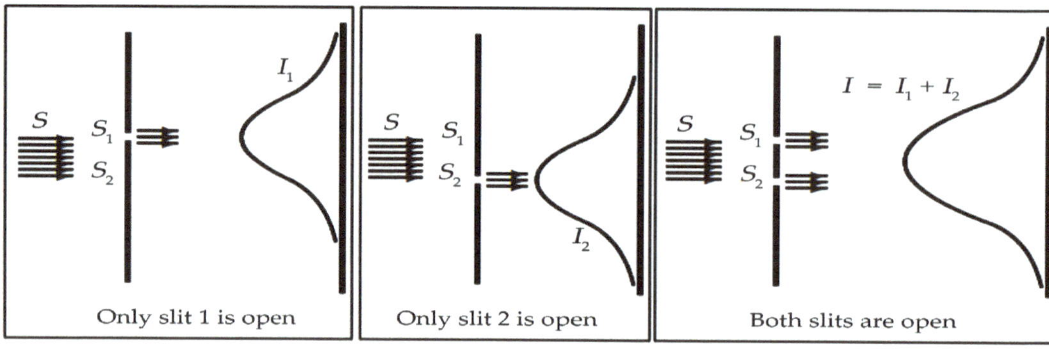

Figure 1.6. Illustration of double-slit experiment in which S is a source of bullet streams.

Source: Milton Cha, Creative Commons License.

Where; S is a source of waves

Figure 1.7 shows that S has become a source of several types of waves, such as light and water waves. Let I_1 be the intensity measured on the screen when only S1 is open, and let I_2 denote the intensity measured on the screen when only S2 is open. It is important to note that the intensity of a wave is directly proportional to the square of its amplitude, which can be represented by the height of the water or the electric field. Mathematically, the intensity of a wave can be calculated using the formulas $I_1 = |\varphi_1|^2$, $I_2 = |\varphi_2|^2$. A wave is mathematically described by the complex function O. The superposition of the intensities viewed on the screen when both slits are open exhibits an interference pattern (Rosenblum & Havas, 1972), hence it cannot be equal to the total of I_1 and I_2. The overall magnitude φ is determined by the combination of φ1 and φ2. Therefore, the overall strength is determined by the magnitudes, rather than the strengths (Perrard et al., 2014).

$$
\begin{aligned}
I = |\psi_1 + \psi_2|^2 &= |\psi_1|^2 + |\psi_2|^2 + \left(\psi_1^* \psi_2 + \psi_2^* \psi_1\right) = I_1 + I_2 + 2\mathrm{Re}(\psi_1^* \psi_2) \\
&= I_1 + I_2 + 2\sqrt{I_1 I_2} \cos \delta,
\end{aligned}
$$

Where δ is the oscillating term $2\sqrt{I_1 I_2} \cos \delta$ that causes the interference pattern (Figure 1.7) and is the phase difference between φ1 and φ2.

Since the intensity distribution is influenced by the phase δ, which remains unknown when only one slit is open (although it can be calculated based on the separation of the slits or the observed intensities I_1, I_2, and I) (Frenkel, 1990), it is not possible to estimate it only from I_1, I_2.

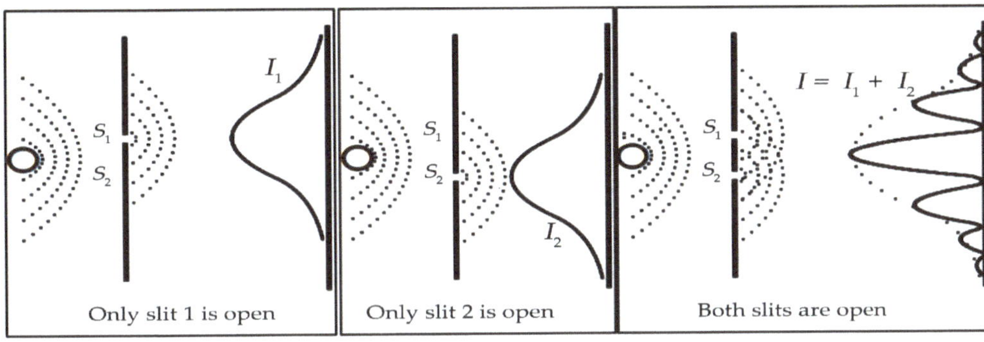

Figure 1.7. Illustration of double-slit experiment in which S is wave's source.

Source: Milton Cha, Creative Commons License.

1.4.2. Quantum View of Waves and Particles

Let's now investigate the double-slit experiment with electrons, or more precisely, with quantum matter particles. Three different experiments with a source (S) that releases an electron stream are shown in Figure 1.8 (Chang, 2021).

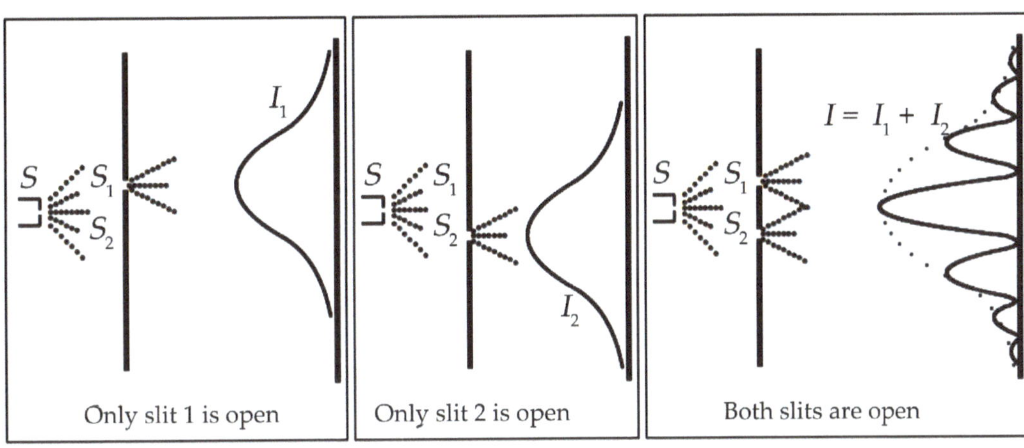

Figure 1.8. Illustration of double-slit experiment when S is an electron source.

Source: Felix Chen, Creative Commons License.

Initially, with S1 closed, subsequently with S2 closed, and lastly with both slits open. The electron distributions on the screen in the first two examples are smooth, and the total of these distributions also exhibits smoothness, resulting in a bell-shaped curve resembling the one found for classical particles (Figure 1.6) (Vigier, 1994).

However, when both slits are open, there is a noticeable and rapid fluctuation in the distribution, resulting in an interesting interference pattern. Electrons are discrete, yet they behave strangely when they interact with one another. This suggests that every electron somehow travels through both slits simultaneously! One can ask how an electron can appear to pass through both slits at the same time if it cannot be divided.

Note that there is no correlation between the interference pattern and the intensity of the electron beam (Mannila et al., 2001).

Experiments were conducted using extremely weak beams, where electrons were sent individually, ensuring that each electron reached the screen before the next one was sent. In this scenario, when both slits are open and given enough time for a significant number of impacts to accumulate on the screen, the interference pattern reappears (Rashkovskiy, 2016).

Now, the critical thing is to identify the slit that the electron passed through. Observing the electrons as they emerge from the slits can be used as an experiment to provide an answer to this question. As seen in Figure 1.9, it involves putting a powerful light source behind the wall that has the slits. Geiger counters are positioned all over the screen so that we can hear a click from the counter each time an electron touches the screen. Every time an electron travels through one of the slits on its route to the counter, it scatters light toward our sight because electric charges scatter light (Bush, 2010).

Whenever a click is heard on the counter, a flash is observed near either S1 or S2, but never simultaneously near both. Upon conducting measurements with both slits unobstructed, it is observed that the distribution closely resembles that of classical projectiles, as shown in Figure 1.6. Notably, the characteristic interference pattern is no longer present. However, when the light source is deactivated, the interference pattern comes back (Cheong & Song, 2014).

Based on this experiment, it can be concluded that the simple act of observing the electrons has a significant impact on their distribution on the screen. Electrons possess a high level of fragility, as their behavior is altered when subjected to observation. This principle in quantum mechanics explains that measurements can disrupt the states of tiny objects. One could consider reducing the intensity of the light source to a level that does not disrupt the electrons (Auletta & Tarozzi, 2004).

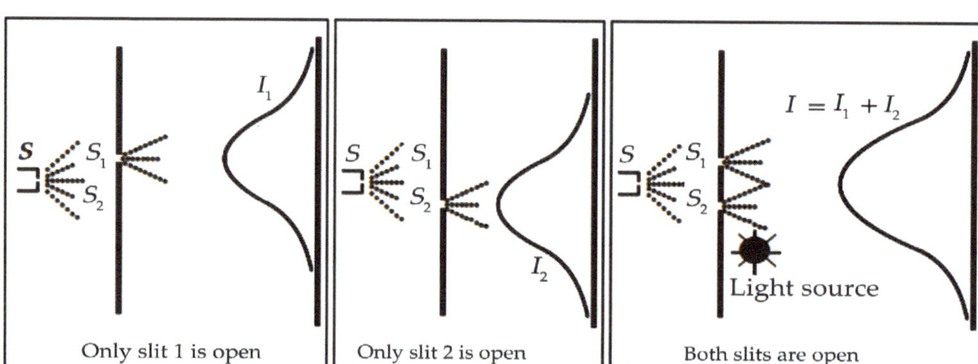

Figure 1.9. Illustration of double-slit experiment when S is a source of electrons. A light source is positioned behind the wall that contains S_1 and S_2.

Source: Felix Chen, Creative Commons License.

It has been observed that the intensity of the scattered light remains constant when electrons pass by, although the flashes occur occasionally. At lower levels of brightness, there is a discrepancy where we detect the presence of electrons through the sound of a click, but do not observe any visible flash. When the brightness levels are reduced even further, a significant number of electrons go undetected (Holland, 2020).

In this scenario, it can be deduced that certain electrons managed to pass through the slits undetected, as there was a lack of photons present at the precise moment to capture their movement. This process is crucial as it provides confirmation that light possesses particle properties. It demonstrates that light arrives at the screen in discrete units known as photons (Bøe & Viefers, 2023).

Two distribution profiles are generated from this experiment with a faint light source. One profile represents the electrons that were observed, while the second profile represents the electrons that were not observed but were detected by the counter. The initial distribution displays no interference, resembling the behavior of classical projectiles. Nevertheless, the resulting distribution displays an interference pattern. This phenomenon arises from the interference exhibited by electrons when they are not under observation. When the electron is not observed, it remains undisturbed by any photons. Nevertheless, when observed, a photon creates a disturbance (Olsen, 2002).

It is impossible to determine which slit each electron that exhibits interference passed through. A new fundamental idea is presented by this experimental result: the microphysical world is indeterministic. In contrast to classical physics, which allows us to precisely track particles along their paths, we are unable to track or identify the journey of a microscopic particle. Technically, it is not possible to track the particle's motion in such detail (Hendry, 1980).

These findings motivated Heisenberg to propose the uncertainty principle, which asserts that it is not feasible to construct a device that can identify the specific slit through which an electron passes without significantly perturbing the electron and thereby obliterating the interference pattern (we will revisit this principle at a later point). The double-slit experiment reveals that electrons have characteristics of both particles and waves, as evidenced by the observed interference pattern (Chowdhury et al., 2021).

When electrons are individually viewed or detected, they exhibit particle-like behavior. However, when they are detected collectively over several measurements (resulting in a distribution of detected electrons), they exhibit wave-like properties with a wavelength $\lambda = h/p$ and show an interference pattern (Dirac, 1932).

1.4.3. Wave-Particle Duality

Thus far, several experimental results have been discussed regarding photoelectric, blackbody radiation, and Compton Effect, Thomson, Davisson–Germer, pair production, and double-slit experiments. These findings demonstrate that electrons, photons, and other microscopic particles behave differently from classical waves and particles. These results suggest that nature can exhibit both wave and particle behavior at the tiny level (Afshar et al., 2007).

The appropriate framework for reconciling matter's particle and wave properties is provided by quantum physics. Quantum mechanics can provide a simultaneous knowledge of the behavior of material particles, such electrons, and the wave-like characteristics of microscopic systems by using a wave function to describe them. The idea is to combine wave-based theories of matter with energy quantization. It describes the properties of the same material particle using both wave and particle representations (Dittel et al., 2021).

Thus, our everyday understanding of particles or waves is insufficient to describe tiny systems. These two ideas do not strictly apply to the microphysical world, even though they contradict each other in the macroscopic domain. At the tiny scale, the idea that a wave cannot act like a particle and vice versa is no longer true. A quantum system is neither a pure wave nor particle in its true sense of the word (Butto, 2020).

The dual nature of a quantum system, characterized by both particle and wave properties, becomes apparent only when it is observed using methods that can penetrate its nature. However, any such observation would inevitably disrupt the initial state of the quantum system. For example, simply observing an electron would cause it to deviate from its original orbit. The electron can exhibit either particle or wave characteristics, depending on the sort of equipment used for observation (Jia et al., 2014).

As demonstrated by the double-slit experiment, examining the particle nature of the electron requires blocking one slit or introducing an observational apparatus. On the other hand, to study its wave characteristics, both slits must remain open without any interference from observational tools. It is interesting to observe that the electron possesses both "grain" and "wave" characteristics. By adjusting the probing tool, we can selectively suppress one aspect of the electron while preserving the other (Pons et al., 2012).

A quantum system experiment that seeks to isolate the particle properties fails to provide any information about its wave properties, and vice versa. During Compton scattering, the particle properties of an electron are the only observable aspects. Nevertheless, when we incorporate it into a diffraction experiment like Davisson-Germer, Thomson, or the double-slit experiment, we exclusively perceive its wave characteristics. Measuring the qualities of a quantum system will result in the destruction of its wave properties, and the opposite is also true. Every measurement yield either one characteristic or the other, but never both simultaneously. We can notice either the wave property or the particle property, but not both simultaneously (Sperling et al., 2019).

As a result, microscopic systems are both pure waves and pure particles. As proposed by Bohr, the particle and wave manifestations are just complimentary rather than incompatible or excluding of one another. When it comes to explaining the actual nature of tiny systems, these ideas work well together. For a comprehensive explanation of quantum systems, waves and particles, complementary properties of microscopic matter are equally significant. This is where the complementarity principle gets its essence. We have seen that when we apply or impose on quantum systems the inflexible concept of either/or (i.e., either a particle or a wave), we run short of reality (Agop & Nica, 2000).

1.5. INDETERMINISTIC NATURE OF MICROPHYSICAL WORLD

First, let's discuss the two significant experimental results that were previously mentioned. On one hand, interference patterns are indeed produced by minute material particles, as demonstrated by the Davisson-Germer and double-slit investigations. We have shown that it is essential to characterize microscopic particles in terms of waves to explain the interference pattern (Bradley, 1962).

Waves exhibit a lack of localization in space. When it comes to describing microscopic particles, we often must sacrifice accuracy because waves can only provide a probabilistic account at best. However, in the double-slit experiment, it has been observed that tracking the movement of individual electrons is not feasible. There is currently no experimental apparatus capable of determining which slit a specific electron has passed through. The inability to predict individual events goes against a fundamental principle of classical physics: determinism or predictability. Heisenberg proposed that the microphysical universe is unexpected based on the experimental results and Born established the idea of probability as a means of understanding quantum mechanics (Hobbs, 1991).

1.5.1. Heisenberg's Uncertainty Principle

According to classical physics, forces acting on a physical system and its initial circumstances can be used to exactly predict how the system will behave in the future. Newton's second law allows for the unique determination of the position , velocity , provided that the initial coordinates , velocity , and all forces acting on the particle are known. Classical physics is wholly deterministic in nature (Barukčić, 2016).

Does the microphysical universe adhere to this deterministic perspective as well? In quantum physics, a particle is not bound to a specific location in space but rather is distributed generally over space according to its wave function. This contrasts with classical particles, which are essentially point particles (Schlosshauer et al., 2013).

Moreover, the double-slit experiment has conclusively shown that it is impossible to determine the precise location of the electron's passage through the slit without causing interference. Hence, the traditional concepts of a particle's distinct trajectory, precise location, and definite momentum become irrelevant when examining phenomena at the microscopic level. The following is a fundamental explanation of Heisenberg's uncertainty principle (Marciak & Kozlowski, 2013).

Heisenberg's uncertainty principle, in its original version, says that if a particle's x-component of momentum is measured with an uncertainty of Δp_x, then its x-position cannot be determined more precisely than $\Delta x = h/(2\Delta p_x)$. The uncertainty relations for momentum and position can be expressed in the following three-dimensional form (Dalla et al., 2007).

$$\Delta x \Delta p_x \geq \frac{\hbar}{2}, \qquad \Delta y \Delta p_y \geq \frac{\hbar}{2}, \qquad \Delta z \Delta p_z \geq \frac{\hbar}{2}.$$

This principle highlights the limitation of simultaneously measuring the momentum and position of a particle with arbitrary accuracy. While it is feasible to accurately measure either the momentum or position, it is not possible to do so for both observables at the same time. It is impossible to locate the exact position of a tiny particle without imparting a significant amount of momentum to it. Measuring the position of an object in quantum physics is a delicate task that inevitably alters its momentum. Passive measures are impossible in this case. To understand this concept, let's analyze the procedure of determining the location of a large thing, such as an automobile, and the location of a small system, such an electron inside an atom (Ozawa, 2003).

To determine the position of a macroscopic object, one can simply observe it. The light that interacts with the object and is reflected to the detector, whether it be our eyes or a measuring device, does not have any noticeable impact on the object's motion. Alternatively, when determining the position of an electron within an atom, it is necessary to use radiation with an extremely short wavelength, comparable to the size of the atom. The energy of this radiation is significant enough to greatly alter the momentum of the electron. The electron's motion is greatly influenced by the act of observation alone; in fact, it can be entirely removed from its orbit. Thus, it is difficult to accurately estimate both position and momentum at the same time (Atkinson & Peijnenburg, 2022).

A confined particle would have a very small wavelength since its wave function would become zero everywhere else. This particle will have a relatively high momentum based on de Broglie's relation, $p=h\lambda$. Formally stated, this indicates that there will be complete uncertainty regarding a particle's momentum ($\Delta px \to \infty$) if it is precisely localized ($\Delta x \to 0$). In summary, we are forced to accept limitations on our capacity to measure any two complimentary variables simultaneously because waves represent all quantum processes (Maor, 2009).

Heisenberg's uncertainty principle can be applied to any pair of dynamical variables that are complementary or canonically conjugate. It asserts that creating an experiment that can arbitrarily accurately assess two complementary variables at the same time is impossible. The theory of quantum mechanics would break down if such a measurement were ever made. Time and energy are two complementary factors. The time-energy uncertainty connection must be followed in their simultaneous measurement (Nikjoo et al., 2008):

$$\Delta E \Delta t \geq \frac{\hbar}{2}.$$

According to the above relation, the difference in the observed energies, represented as ΔE, between two energy measurements of a system performed at different times must be less than $h/\Delta t$. A large, elapsed period between the two observations will lead to a negligible energy difference. This behavior can be explained by the fact that the system is perturbed following the first measurement and takes a long time to recover to its initial, undisturbed condition. This formula is highly useful for analyzing decay processes since it clearly specifies the relationship between the excited states' energy range and average lifespan (Werner & Farrelly, 2019).

Quantum mechanics, in contrast to traditional physics, is a theory that is completely indeterministic. When attempting to figure out the position or momentum of an electron, it is fundamentally impossible to obtain a definitive answer. However, the only response that is currently accessible is one that is based on probability. According to principle of uncertainty, when exact position of a quantum system is known, its momentum becomes wholly unpredictable. In this specific setting, the uncertainty principle has unquestionably posed a threat to a fundamental premise of classical physics: the predictable and deterministic character of Newtonian mechanics (Yu et al., 2020).

1.6. ATOMIC TRANSITIONS AND SPECTROSCOPY

In addition to its inability to clarify blackbody radiation, classical physics also falls short in explaining phenomena such as the Compton Effect, photoelectric effect, pair creation effect, wave-particle duality, and several other microscopic-scale phenomena. Here, we examine another domain in which conventional physics fails to apply the atom. Empirical findings demonstrate that atoms exist as stable, bonded entities with distinct energy levels. According to classical physics, it is stated that any confined system must possess a continuous range of energy levels (Kazakov et al., 2017).

1.6.1. Planetary Model of Atom by Rutherford

In 1911, Rutherford made an important breakthrough when he identified the atomic nucleus, leading to the development of a model to describe the properties of the atom. Rutherford believed that electrons in an atom revolved around the nucleus, a positively charged massive center, having been inspired by the planets' revolving motion around the sun (Kragh, 2021).

It quickly emerged that this model had two major flaws that seriously undermine classical physics: (a) Atoms exhibit instability, and (b) Atoms emit energy constantly across a broad spectrum of frequencies. Applying Maxwell's electromagnetic theory to Rutherford's model reveals first shortcoming: the electron accelerates as it orbits the nucleus and releases energy as a result. As a result, it must lose energy (Kuehn & Kuehn, 2016).

After that, the orbit's radius should continuously shrink (spiral motion) until the electron crashes onto the nucleus; this usually takes 10^{-8} s. Secondly, the orbiting frequency of an electron increases constantly as its orbit shrinks because the radiated energy frequency and orbiting frequency are the same. As a result, the atom should produce radiation with a continuous spectrum. Given that atoms are stable and emit energy within specific frequency ranges, these two conclusions are in direct opposition to the results of the experiment (Heilbron, 1985).

1.6.2. Model of Hydrogen Atom by Bohr

In 1913, Bohr proposed a model that integrates Planck's quantum hypothesis, Rutherford's planetary model and Einstein's photon concept. This model provides an exact explanation for the observable spectrum of the hydrogen atom and convincingly explains its stability. Bohr postulated, similar to Rutherford's model that the electron of each atom revolves in an orbit around the nucleus due to the electrostatic attraction of the nucleus. Classical mechanics permits both circular and elliptical orbits. Bohr simplified the analysis by focusing only on circular orbits and made a number of arbitrary assumptions that contradict classical physics (Olszewski, 2016). However, these assumptions have shown to be quite effective in describing numerous characteristics of the hydrogen atom:

i. In contrast to classical mechanics, where a range of orbits is feasible, only a limited number of circular stable orbits, known as stationary states, are permitted. Atoms can only exist in specific stable states characterized by distinct energies, such as E1, E2, E3, and so on.

ii. The orbits that are permitted (stationary) are those for which the electron's orbital angular momentum is an integer multiple of h (h=ħ/2π).

  a. L=nh

The relationship is referred to as the Bohr quantization rule for angular momentum (Hughes, 1990).

iii. Electrons in fixed orbits do not emit electromagnetic energy. The emission or absorption of radiation occurs exclusively when an electron transitions from one permissible orbit to another. The emission of radiation occurs when an electron transitions from one orbit with energy En to another orbit with energy Em. This emission is carried out by a photon with energy proportional to the transition (Svidzinsky et al., 2014).

$$hv=E_n - E_m$$

An atom can release (or absorb) radiation by the transition of an electron to a lower (or higher) orbit (Alnes et al., 2007).

ACTIVITY 1.1

A blackbody with a temperature of 3000 K emits radiation with a peak wavelength of around 966 nm. Apply Planck's law and Wien's displacement law to calculate the wavelength at which a blackbody at a temperature of 6000 K emits the most intense radiation.

SUMMARY

- This chapter provides a comprehensive overview of the key discoveries and concepts that led to the development of quantum mechanics. It begins with a historical context, highlighting the limitations of classical physics in explaining certain phenomena, which necessitated the creation of a new theoretical framework.
- The chapter first delves into the particle nature of radiation, starting with blackbody radiation. Max Planck introduced the idea of quantized energy levels to address the issue of the ultraviolet catastrophe. The section also covers the photoelectric effect, a phenomenon explained by Albert Einstein's hypothesis that light consists of discrete energy units called photons. This theory explains how electrons are emitted from a metal surface when exposed to light.
- The de Broglie hypothesis is then discussed, which explores the wave-like properties of particles, suggesting that particles such as electrons exhibit wave-like behavior. This idea was pivotal in developing the wave-particle duality concept, a fundamental principle of quantum mechanics.
- The chapter continues by comparing classical and quantum perspectives on particles and waves. In classical physics, particles and waves are distinct entities with different characteristics. Quantum mechanics introduces the concept of wave-particle duality, where particles can display both wave-like and particle-like properties depending on the experimental conditions. The Heisenberg Uncertainty Principle, a key concept in quantum mechanics, asserts that it is impossible to precisely determine both a particle's momentum and position simultaneously, highlighting the inherent limitations in measuring quantum systems and the probabilistic nature of quantum mechanics.
- The chapter also covers atomic transitions and spectroscopy, beginning with the Rutherford planetary model, which depicted electrons orbiting a nucleus but failed to explain atomic stability. The Bohr model of the hydrogen atom is then examined in detail. This model introduced quantized electron orbits and successfully described atomic spectra through electron transitions between these orbits.

REVIEW QUESTIONS

1. What were the limitations of classical physics that led to the development of Quantum Mechanics?
2. Explain the significance of the photoelectric effect and how Einstein's interpretation challenged classical wave theory?
3. Describe de Broglie's hypothesis and its implications for the nature of matter?
4. Compare and contrast the classical and quantum views of particles and waves?

5. What are the fundamental principles that differentiate Quantum Mechanics from classical physics?

MULTIPLE CHOICE QUESTIONS

1. Which of the following phenomena could not be explained by classical physics and led to the development of Quantum Mechanics

a. Gravitational attraction

b. Blackbody radiation

c. Projectile motion

d. Fluid dynamics

2. Who introduced the concept of quantized energy levels to resolve the ultraviolet catastrophe?

a. Albert Einstein

b. Niels Bohr

c. Max Planck

d. Erwin Schrödinger

3. What is the significance of the photoelectric effect in Quantum Mechanics?

a. It demonstrates the wave nature of light.

b. It shows that energy levels in atoms are continuous.

c. It provides evidence for the particle nature of light.

d. It explains the formation of atomic spectra.

4. Which experiment is most closely associated with demonstrating wave-particle duality

a. Michelson-Morley experiment

b. Double-slit experiment

c. Rutherford gold foil experiment

d. Cavendish experiment

5. The Heisenberg Uncertainty Principle states that it is impossible to simultaneously know which two properties of a particle with arbitrary precision?

a. Energy and time

b. Position and momentum

c. Mass and charge

d. Velocity and acceleration

Answers to Multiple Questions

1. (b); 2. (c); 3. (c); 4. (b); 5. (b)

REFERENCES

1. Aatrokoski, J., Ade, P. A., Aghanim, N., Aller, H. D., Aller, M. F., Angelakis, E., & Naselsky, P. (2011). Planck early results. XV. Spectral energy distributions and radio continuum spectra of northern extragalactic radio sources. Astronomy & Astrophysics, 536, A15.
2. Afshar, S. S., Flores, E., McDonald, K. F., & Knoesel, E. (2007). Paradox in wave-particle duality. Foundations of Physics, 37(3), 295–305.
3. Agop, M., & Nica, P. (2000). The wave-particle duality in the Weyl-Dirac theory. Classical and Quantum Gravity, 17(18), 3627–3634.
4. Alnes, H., Ravndal, F., & Wehus, I. K. (2007). Black-body radiation with extra dimensions. Journal of Physics A: Mathematical and Theoretical, 40(47), 14309.
5. Atkinson, D., & Peijnenburg, J. (2022). How certain is Heisenberg's uncertainty principle? HOPOS: The Journal of the International Society for the History of Philosophy of Science, 12(1), 1–21.
6. Auletta, G., & Tarozzi, G. (2004). On the physical reality of quantum waves. Foundations of Physics, 34(11), 1675–1694.
7. Baracca, A., Bergia, S., & Del Santo, F. (2017). The origins of the research on the foundations of quantum mechanics (and other critical activities) in Italy during the 1970s. Studies in History and Philosophy of Science Part B: Studies in History and Philosophy of Modern Physics, 57, 66–79.
8. Barukčić, I. (2016). Anti Heisenberg—The end of Heisenberg's uncertainty principle. Journal of Applied Mathematics and Physics, 4(5), 881–887.
9. Blum, A. S., & Jähnert, M. (2022). The birth of quantum mechanics from the spirit of radiation theory. Studies in History and Philosophy of Science, 91, 125–147.
10. Bøe, M. V., & Viefers, S. (2023). Secondary and university students' descriptions of quantum uncertainty and the wave nature of quantum particles. Science & Education, 32(2), 297–326.
11. Bradley, R. D. (1962). Determinism or indeterminism in microphysics. The British Journal for the Philosophy of Science, 13(51), 193–215.
12. Brida, G., Cagliero, E., Falzetta, G., Genovese, M., Gramegna, M., & Novero, C. (2002). Experimental realization of a first test of de Broglie–Bohm theory. Journal of Physics B: Atomic, Molecular and Optical Physics, 35(22), 4751–4762.
13. Bub, J. (2000). Quantum mechanics as a principal theory. Studies in History and Philosophy of Science Part B: Studies in History and Philosophy of Modern Physics, 31(1), 75–94.
14. Bunge, M. (1956). Survey of the interpretations of quantum mechanics. American Journal of Physics, 24(4), 272–286.
15. Bush, J. W. (2010). Quantum mechanics writ large. Proceedings of the National Academy of Sciences, 107(41), 17455–17456.
16. Butto, N. (2020). A new theory on electron wave-particle duality. Journal of High Energy Physics, Gravitation and Cosmology, 6(4), 567–578.

17. Castellani, L. (2019). History operators in quantum mechanics. International Journal of Quantum Information, 17(08), 1941001.

18. Catillon, P., Cue, N., Gaillard, M. J., Genre, R., Gouanère, M., Kirsch, R. G., & Spighel, M. (2008). A search for the de Broglie particle internal clock by means of electron channeling. Foundations of Physics, 38(5), 659–664.

19. Chang, D. C. (2021). Review on the physical basis of wave–Particle duality: Conceptual connection between quantum mechanics and the Maxwell theory. Modern Physics Letters B, 35(13), 2130004.

20. Cheong, Y. W., & Song, J. (2014). Different levels of the meaning of wave-particle duality and a suspensive perspective on the interpretation of quantum theory. Science & Education, 23(8), 1011–1030.

21. Chowdhury, P., Pati, A. K., & Chen, J. L. (2021). Wave and particle properties can be spatially separated in a quantum entity. Photonics Research, 9(7), 1379–1383.

22. Christensen, O. (1976). Quantum efficiency of the internal photoelectric effect in silicon and germanium. Journal of Applied Physics, 47(2), 689–695.

23. Compton, K. T., & Richardson, O. W. (1913). The photoelectric effect—II. The London, Edinburgh, and Dublin Philosophical Magazine and Journal of Science, 26(154), 549–567.

24. Cooke, W. E., & Gallagher, T. F. (1980). Effects of blackbody radiation on highly excited atoms. Physical Review A, 21(2), 588–593.

25. Dalla Chiara, M. L., Giuntini, R., & Rédei, M. (2007). The history of quantum logic. In The Many Valued and Nonmonotonic Turn in Logic (Vol. 8, pp. 205–283).

26. Darrigol, O. (2009). A simplified genesis of quantum mechanics. Studies in History and Philosophy of Science Part B: Studies in History and Philosophy of Modern Physics, 40(2), 151–166.

27. David, F. (2015). The formalisms of quantum mechanics. In Lecture Notes in Physics (Vol. 893, pp. 42–43).

28. De Ninno, G., Wätzel, J., Ribič, P. R., Allaria, E., Coreno, M., Danailov, M. B., & Berakdar, J. (2020). Photoelectric effect with a twist. Nature Photonics, 14(9), 554–558.

29. DiBenedetto, M. H., Ouellette, N. T., & Koseff, J. R. (2018). Transport of anisotropic particles under waves. Journal of Fluid Mechanics, 837, 320–340.

30. Dirac, P. A. M. (1932). Relativistic quantum mechanics. Proceedings of the Royal Society of London; Series A, Containing Papers of a Mathematical and Physical Character, 136(829), 453–464.

31. Dittel, C., Dufour, G., Weihs, G., & Buchleitner, A. (2021). Wave-particle duality of many-body quantum states. Physical Review X, 11(3), 031041.

32. Dougal, R. C. (1976). The presentation of the Planck radiation formula (tutorial). Physics Education, 11(6), 438–441.

33. Dunbar, R. C., & McMahon, T. B. (1998). Activation of unimolecular reactions by ambient blackbody radiation. Science, 279(5348), 194–197.

34. Dürr, S., Nonn, T., & Rempe, G. (1998). Origin of quantum-mechanical complementarity probed by a 'which-way' experiment in an atom interferometer. Nature, 395(6697), 33–37.
35. Eliezer, C. J. (1948). On the classical theory of particles. Proceedings of the Royal Society of London. Series A. Mathematical and Physical Sciences, 194(1039), 543–555.
36. El'yashevich, M. A. (1977). From the origin of quantum concepts to the establishment of quantum mechanics. Soviet Physics Uspekhi, 20(8), 656–677.
37. Fano, U., McVoy, K. W., & Albers, J. R. (1959). Sauter theory of the photoelectric effect. Physical Review, 116(5), 1147–1161.
38. Field, J. H. (2013). Description of diffraction-grating experiments for photons and electrons in Feynman's spacetime formulation of quantum mechanics: The quantum origins of classical wave theories of light and massive particles. European Journal of Physics, 34(6), 1507–1518.
39. Frenkel, A. (1990). Spontaneous localizations of the wave function and classical behavior. Foundations of Physics, 20(2), 159–188.
40. Frieden, B. R., & Soffer, B. H. (2009). De Broglie's wave hypothesis from Fisher information. Physica A: Statistical Mechanics and its Applications, 388(7), 1315–1330.
41. Gallagher, T. F., & Cooke, W. E. (1979). Interactions of blackbody radiation with atoms. Physical Review Letters, 42(13), 835–838.
42. Garritz, A. (2013). Teaching the philosophical interpretations of quantum mechanics and quantum chemistry through controversies. Science & Education, 22, 1787–1807.
43. Gell-Mann, M., & Hartle, J. B. (2012). Decoherent histories quantum mechanics with one real fine-grained history. Physical Review A: Atomic, Molecular, and Optical Physics, 85(6), 062120.
44. Griffiths, R. B. (1984). Consistent histories and the interpretation of quantum mechanics. Journal of Statistical Physics, 36(3–4), 219–272.
45. Güémez, J., & Fiolhais, M. (2018). Relativistic description of the photoelectric effect. American Journal of Physics, 86(11), 825–830.
46. Halim, A., Soewarno, S., Elmi, E., Zainuddin, Z., Huda, I., & Irwandi, I. (2020). The impact of the e-learning module on remediation of misconceptions in modern physics courses. Jurnal Penelitian & Pengembangan Pendidikan Fisika, 6(2), 203–216.
47. Haslinger, P., Jaffe, M., Xu, V., Schwartz, O., Sonnleitner, M., Ritsch-Marte, M., & Müller, H. (2018). Attractive force on atoms due to blackbody radiation. Nature Physics, 14(3), 257–260.
48. Heilbron, J. L. (1985). Bohr's first theories of the atom. Physics Today, 38(10), 28–36.
49. Hendry, J. (1980). The development of attitudes to the wave-particle duality of light and quantum theory, 1900–1920. Annals of Science, 37(1), 59–79.
50. Hilgevoord, J. (2005). Time in quantum mechanics: A story of confusion. Studies in History and Philosophy of Science Part B: Studies in History and Philosophy of Modern Physics, 36(1), 29–60.

51. Hobbs, J. (1991). Chaos and indeterminism. Canadian Journal of Philosophy, 21(2), 141–164.

52. Holland, P. (2020). Uniting the wave and the particle in quantum mechanics. Quantum Studies: Mathematics and Foundations, 7, 155–178.

53. Hughes, R. I. G. (1990). The Bohr atom, models, and realism. Philosophical Topics, 18(2), 71–84.

54. Hulme, H. R. (1931). The photoelectric effect for γ-rays. Proceedings of the Royal Society of London. Series A, Containing Papers of a Mathematical and Physical Character, 133(822), 381–406.

55. Hund, F. (1977). Early history of the quantum mechanical treatment of the chemical bond. Angewandte Chemie International Edition in English, 16(2), 87–91.

56. Ireson, G. (2000). A brief history of quantum phenomena. Physics Education, 35(6), 381–386.

57. Jia, A. A., Huang, J. H., Zhang, T. C., & Zhu, S. Y. (2014). Influence of losses on the wave-particle duality. Physical Review A, 89(4), 042103.

58. Kaniadakis, G. (2002). Statistical origin of quantum mechanics. Physica A: Statistical Mechanics and its Applications, 307(1–2), 172–184.

59. Kazakov, V. V., Kazakov, V. G., Kovalev, V. S., Meshkov, O. I., & Yatsenko, A. S. (2017). Electronic structure of atoms: Atomic spectroscopy information system. Physica Scripta, 92(10), 105002.

60. Koch, M., Pielsticker, S., & Kneer, R. (2024). Biomass particle-radiation interaction and the effect of shape and structure simplifications. Fuel, 361, 130657.

61. Kovačević, M. S., & Djordjevich, A. (2006). A mechanical analogy for the photoelectric effect. Physics Education, 41(6), 551–556.

62. Kragh, H. (2021). Chemical and other aspects of Rutherford's nuclear atom. Journal of the Royal Society of New Zealand, 51(3–4), 513–527.

63. Kuehn, K., & Kuehn, K. (2016). Bohr's atomic model. In A Student's Guide Through the Great Physics Texts: Volume IV: Heat, Atoms and Quanta (pp. 377–391).

64. Mannila, K., Koponen, I. T., & Niskanen, J. A. (2001). Building a picture of students' conceptions of wave-and particle-like properties of quantum entities. European Journal of Physics, 23(1), 45–53.

65. Maor, E. (2009). How I got to understand Heisenberg's uncertainty principle. Math Horizons, 16(4), 5–33.

66. Marciak-Kozlowska, J., & Kozlowski, M. (2013). Heisenberg's uncertainty principle and human brain. NeuroQuantology, 11(1), 11–18.

67. Maslovski, S. I., Simovski, C. R., & Tretyakov, S. A. (2016). Overcoming black body radiation limit in free space: Metamaterial super emitter. New Journal of Physics, 18(1), 013034.

68. McKagan, S. B., Perkins, K. K., Dubson, M., Malley, C., Reid, S., LeMaster, R., & Wieman, C. E. (2008). Developing and researching PhET simulations for teaching quantum mechanics. American Journal of Physics, 76(4), 406–417.

69. Miaja-Avila, L., Lei, C., Aeschlimann, M., Gland, J. L., Murnane, M. M., Kapteyn, H. C., & Saathoff, G. (2006). Laser-assisted photoelectric effect from surfaces. Physical Review Letters, 97(11), 113604.
70. Muller, F. A. (1997). The equivalence myth of quantum mechanics—Part I. Studies in History and Philosophy of Science Part B: Studies in History and Philosophy of Modern Physics, 28(1), 35–61.
71. Narayanaswamy, A., Shen, S., Hu, L., Chen, X., & Chen, G. (2009). Breakdown of the Planck blackbody radiation law at nanoscale gaps. Applied Physics A, 96, 357–362.
72. Nikjoo, H., Uehara, S., Emfietzoglou, D., & Brahme, A. (2008). Heavy charged particles in radiation biology and biophysics. New Journal of Physics, 10(7), 075006.
73. Olsen, R. V. (2002). Introducing quantum mechanics in the upper secondary school: A study in Norway. International Journal of Science Education, 24(6), 565–574.
74. Olszewski, S. (2016). The Bohr model of the hydrogen atom revisited. Reviews in Theoretical Science, 4(4), 336–352.
75. Ossiander, M., Riemensberger, J., Neppl, S., Mittermair, M., Schäffer, M., Duensing, A., & Kienberger, R. (2018). Absolute timing of the photoelectric effect. Nature, 561(7723), 374–377.
76. Ozawa, M. (2003). Universally valid reformulation of the Heisenberg uncertainty principle on noise and disturbance in measurement. Physical Review A, 67(4), 042105.
77. Perrard, S., Labousse, M., Miskin, M., Fort, E., & Couder, Y. (2014). Self-organization into quantized eigenstates of a classical wave-driven particle. Nature Communications, 5(1), 3219.
78. Phillips, M. H., Frankel, K. A., Lyman, J. T., Fabrikant, J. I., & Levy, R. P. (1990). Comparison of different radiation types and irradiation geometries in stereotactic radiosurgery. International Journal of Radiation Oncology Biology Physics, 18(1), 211–220.
79. Planck Collaboration, Arnaud, M., Ashdown, M., Atrio-Barandela, F., Aumont, J., Baccigalupi, C., & Morgante, G. (2016). Planck intermediate results. Astronomy & Astrophysics, 586, A134.
80. Pons, D. J., Pons, A. D., Pons, A. M., & Pons, A. J. (2012). Wave-particle duality: A conceptual solution from the cordus conjecture. Physics Essays, 25(1), 132–140.
81. Popa, L. A., & Vasile, A. (2008). WMAP five-year constraints on lepton asymmetry and radiation energy density: Implications for Planck. Journal of Cosmology and Astroparticle Physics, 2008(06), 028.
82. Prokhorov, A. V., Hanssen, L. M., & Mekhontsev, S. N. (2009). Calculation of the radiation characteristics of blackbody radiation sources. Experimental Methods in the Physical Sciences, 42, 181–240.
83. Rashkovskiy, S. A. (2016). Quantum mechanics without quanta: The nature of the wave–Particle duality of light. Quantum Studies: Mathematics and Foundations, 3, 147–160.

84. Rosenblum, A., & Havas, P. (1972). Classical theory of particles interacting with electromagnetic and mesonic fields. III. Physical Review D, 6(6), 1522–1532.

85. Saunders, S. (1998). Time, quantum mechanics, and probability. Synthese, 114, 373–404.

86. Schirripa Spagnolo, G., Leccese, F., & Leccisi, M. (2019). LED as transmitter and receiver of light: A simple tool to demonstrate the photoelectric effect. Crystals, 9(10), 531.

87. Schlosshauer, M., Kofler, J., & Zeilinger, A. (2013). A snapshot of foundational attitudes toward quantum mechanics. Studies in History and Philosophy of Science Part B: Studies in History and Philosophy of Modern Physics, 44(3), 222–230.

88. Schmidt-Böcking, H. (2019). The Stern-Gerlach experiment re-examined by an experimenter. Europhysics News, 50(3), 15–19.

89. Smoot, G. F., Gorenstein, M. V., & Muller, R. A. (1977). Detection of anisotropy in the cosmic blackbody radiation. Physical Review Letters, 39(14), 898–901.

90. Sorokin, A. A., Bobashev, S. V., Feigl, T., Tiedtke, K., Wabnitz, H., & Richter, M. (2007). Photoelectric effect at ultrahigh intensities. Physical Review Letters, 99(21), 213002.

91. Spagnolo, G. S., Postiglione, A., & De Angelis, I. (2020). Simple equipment for teaching internal photoelectric effect. Physics Education, 55(5), 055011.

92. Sperling, J., De, S., Nitsche, T., Tiedau, J., Barkhofen, S., Brecht, B., & Silberhorn, C. (2019). Wave-particle duality revisited: Neither wave nor particle. arXiv Preprint. https://arxiv.org/abs/1907.09836 (accessed on 7 September 2024).

93. Stepansky, B. K. (1997). Ambiguity: Aspects of the wave–Particle duality. The British Journal for the History of Science, 30(3), 375–385.

94. Svidzinsky, A., Scully, M., & Herschbach, D. (2014). Bohr's molecular model, a century later. Physics Today, 67(1), 33–39.

95. Taslidere, E. (2016). Development and use of a three-tier diagnostic test to assess high school students' misconceptions about the photoelectric effect. Research in Science & Technological Education, 34(2), 164–186.

96. Thomson, G. P. (1928). Experiments on the diffraction of cathode rays. Proceedings of the Royal Society of London. Series A, Containing Papers of a Mathematical and Physical Character, 117(778), 600–609.

97. Townes, C. H. (2005). Early history of quantum electronics. Journal of Modern Optics, 52(12), 1637–1645.

98. Verde, L., Protopapas, P., & Jimenez, R. (2013). Planck and the local Universe: Quantifying the tension. Physics of the Dark Universe, 2(3), 166–175.

99. Vigier, J. P. (1994). Possible test of the reality of superluminal phase waves and particle phase space motions in the Einstein-de Broglie-Bohm causal stochastic interpretation of quantum mechanics. Foundations of Physics, 24(1), 61–83.

100. Weiss, D. S. (2022). Davisson-Germer Prize in Atomic or Surface Physics: Exciting 1D Bose gases. In APS Division of Atomic, Molecular and Optical Physics Meeting Abstracts (Vol. 2022, pp. A01–001).

101. Werner, R. F., & Farrelly, T. (2019). Uncertainty from Heisenberg to today. Foundations of Physics, 49, 460–491.

102. Wetterich, C. (2010). Quantum particles from classical statistics. Annalen der Physik, 522(11), 807–848.

103. Wheaton, B. R. (1978). Philipp Lenard and the photoelectric effect, 1889–1911. Historical Studies in the Physical Sciences, 9, 299–322.

104. Wiseman, H. M. (2006). From Einstein's theorem to Bell's theorem: A history of quantum non-locality. Contemporary Physics, 47(2), 79–88.

105. Wong, D., Lee, P., Shenghan, G., Xuezhou, W., Qi, H. Y., & Kit, F. S. (2011). The photoelectric effect: Experimental confirmation concerning a widespread misconception in the theory. European Journal of Physics, 32(4), 1059–1071.

106. Yamada, I., Narihara, K., Funaba, H., Minami, T., Hayashi, H., Kohmoto, T., & LHD Experiment Group. (2010). Recent progress of the LHD Thomson scattering system. Fusion Science and Technology, 58(1), 345–351.

107. Yu, H., Chen, X., Zhou, Y., Chen, D., & Zhang, L. (2020). Impact of photoelectric effect on X-ray density logging and its correction. Applied Radiation and Isotopes, 156, 108785.

CHAPTER

2

Postulates of Quantum Mechanics

LEARNING OBJECTIVES

After studying this chapter, you will be able to:

1. Understand the basic postulates of the quantum states.
2. Understand the significance of wave functions and superposition in quantum mechanics.
3. Understand the relationship between observables and Hermitian operators.
4. Understand how measurements affect quantum systems.
5. Explore the significance of expectation values and uncertainty relations.
6. Understand the concept of matter waves and the de Broglie equation.
7. Explore the role of the wave function in quantum mechanics.

KEY TERMS FROM THIS CHAPTER

expectation values	Hermitian
measurements	observables
operators	quantum mechanics
superposition	time evolution
uncertainty principle	wave function

UNIT INTRODUCTION

Quantum mechanics is founded upon several postulates. These postulates are derived from a diverse set of experimental observations, which are rooted in the fundamental physical concepts behind these observations. This chapter provides a formal analysis of these postulates and explains how they might be utilized to derive precise numerical data regarding microphysical systems (Carcassi et al., 2021).

These postulates are not deducible; they arise from experimentation. They embody the essential set of postulates required for the formulation of the theory of quantum mechanics. However, how can one ascertain the veracity of these postulates? Their validity cannot be ascertained directly; only an indirect inferential statement is feasible (Marshman & Singh, 2019). To understand this, one must refer to the theory based on these postulates: if the theory is successful, the postulates will be deemed valid; otherwise, they will be rendered meaningless. The efficacy of quantum theory is not only substantial but also exceptionally remarkable, thereby substantiating its experimental validity. It possesses a highly effective ability to make accurate predictions, both in terms of quality and quantity. This predictive ability has been confirmed by a wide range of investigations. The precise predictive capability of quantum theory provides undeniable proof of the correctness of the fundamental assumptions on which the theory is based (Masanes et al., 2019).

This chapter introduces the fundamental postulates of quantum mechanics, outlining the principles that govern the behavior of quantum systems. It discusses the representation of the state of a system through wave functions, the concept of probability density, and the superposition principle. The chapter further explores the relationship between observables and operators, detailing the measurement process and its impact on quantum states. Additionally, it covers expectation values, complete sets of commuting operators (CSCO), and the implications of Heisenberg's uncertainty principle. Finally, the chapter examines the wave-particle duality manifested in matter waves, providing a comprehensive overview of key quantum mechanical concepts (Faraggi & Matone, 2000).

2.1. THE BASIC POSTULATES OF QUANTUM MECHANICS

In classical mechanics, the location and momentum of a particle at any given time t are the two fundamental dynamical variables that specify its state. All additional physical quantities that are pertinent to the system can be computed based on these two dynamic variables (Paris, 2012). Furthermore, by knowing these variables at a specific time t, one can make predictions about their values at any subsequent time t using equations such as Hamilton's equations .

To comprehend these concepts, postulates are employed to specify their quantum mechanical counterparts (Klyshko, 1998).

These concepts' equivalents in quantum mechanics are delineated by postulates, which allow one to comprehend:

a. The mathematical description of a quantum state at a specific time t.
b. The different physical quantities that are based on this quantum state.
c. To determine the state of a system at any given time t and explain its evolution over time, one must ascertain the system's state at that specific time and subsequently determine its condition at any later time (Omnes, 1988).

The solutions to these inquiries are furnished by the subsequent collection of five postulates (Mielnik, 1974).

2.1.1. Postulate 1: State of a System

The state of a physical system is determined by a state vector in a Hilbert space at each time t. This state vector contains all the necessary information about the system and serves as the basis for extracting that information. A state vector is defined as any superposition of state vectors (Masanes et al., 2013).

2.1.2. Postulate 2: Observables and Operators

For any physically measurable quantity A, which is also known by the term observable or dynamical variable, there is a corresponding linear Hermitian operator . The eigenvectors of this operator constitute a full basis (Gudder & Mullikin, 1973).

2.1.3. Postulate 3: Measurements and Eigenvalues of Operators

The formal representation of measuring an observable A involves applying the operator Â on a state vector . An eigenvalue of the operator which is a real number, is the sole possible outcome of such a measurement (Van den & Iversen, 2013). After measuring the state . and obtaining the result a_n, the system's state is immediately updated to .

$$\hat{A}|\psi(t)\rangle = a_n|\psi_n\rangle, \qquad (1)$$

Where Note: a_n is the component of when projected onto the eigenvector .

2.1.4. Postulate 4: Probabilistic Outcome of Measurements

a. Discrete Spectra: The probability of finding one of the non-degenerate eigenvalues of the operator while measuring the observable A of a system in the state is given by:

$$P_n(a_n) = \frac{|\langle\psi_n|\psi\rangle|^2}{\langle\psi|\psi\rangle} = \frac{|a_n|^2}{\langle\psi|\psi\rangle}, \qquad (2)$$

Where is the eigenstate of with eigenvalue a_n. If the eigenvalue a_n is m-degenerate, becomes:

$$P_n(a_n) = \frac{\sum_{j=1}^{m}|\langle\psi_n^j|\psi\rangle|^2}{\langle\psi|\psi\rangle} = \frac{\sum_{j=1}^{m}|a_n^{(j)}|^2}{\langle\psi|\psi\rangle}. \qquad (3)$$

b. Measurement induces a transformation of the system's state, altering it from to . If the system is already in an eigenstate of , then a measurement of A will provide the associated eigenvalue .

Continuous spectra: The Eqn. (2), which applies to discrete spectra, can be expanded to compute the probability density of receiving a measurement result between and given a system that is initially in a state (Cai et al., 2015).

$$\frac{dP(a)}{da} = \frac{|\psi(a)|^2}{\langle\psi|\psi\rangle} = \frac{|\psi(a)|^2}{\int_{-\infty}^{+\infty}|\psi(a')|^2 da'}; \qquad (4)$$

2.1.5. Postulate 5: Time Evolution of a System

The state vector of a system change over time according to the time-dependent Schrödinger equation:

$$i\hbar\frac{\partial|\psi(t)\rangle}{\partial t} = \hat{H}|\psi(t)\rangle, \qquad (5)$$

Where; is the Hamiltonian operator corresponding to the total energy of the system.

These postulates can be classified into two categories:

a. The initial four descriptions depict the system at a certain moment in time.

b. The fifth demonstrates the temporal evolution of this description (Gillespie, 1976).

The subsequent section will examine the tangible consequences of each of the four postulates. Specifically, it will examine the condition of a quantum system and its explanation, the tangible properties, measurements in quantum mechanics, and ultimately the temporal progression of quantum systems (Hardy et al., 1973).

2.2. THE STATE OF A SYSTEM

In quantum mechanics, a system is described using a mathematical object known as a complex function, which belongs to a Hilbert space. This function, denoted as , contains all the necessary information about the system and allows one to calculate all the relevant physical quantities. The state vector can be represented in two different manners:

a. A wave function in the position space:

b. A momentum wave function in the momentum space:

In quantum mechanics, the state of a one-dimensional particle is described using a complex function , rather than the two real values (x, p) used in conventional physics (Debs & Larson, 1970).

Only wave functions corresponding to physical systems will be utilized.

What are the mathematical criteria that a wave function must fulfill in order to accurately reflect a physical system?

Physically acceptable wave functions must be finite (Primadianto & Lu, 2016), continuous, and single-valued everywhere, together with their first derivatives .

2.2.1. Probability Density

What is the physical significance of a wave function? The meaning of the quantity , is solely determined by the square. As to Born's probabilistic interpretation, the magnitude of squared (Haworth, 2010).

$$P(\vec{r}, t) = |\psi(\vec{r}, t)|^2, \tag{6}$$

The term indicates the probability density of locating the particle at time t within a volume element , which is positioned between and . Hence, the overall chance of locating the system at any given point in space is precisely 1:

$$\int |\psi(\vec{r}, t)|^2 d^3r = \int_{-\infty}^{+\infty} dx \int_{-\infty}^{+\infty} dy \int_{-\infty}^{+\infty} |\psi(\vec{r}, t)|^2 dz = 1. \tag{7}$$

Normalization is defined as satisfying this relation for a wave function . It is worth noting that the function has the physical dimensions of where represents a length (Koliander et al., 2022). Therefore, the magnitude squared of the wavefunction has physical dimensions of .

Did you know?

Particles such as electrons, as well as other forms of matter, have characteristics of both waves and particles. This phenomenon is a fundamental principle in the field of quantum mechanics, referred to as wave-particle duality.

2.2.2. The Superposition Principle

A system's state can be represented not only by a single wave function, but also by a superposition of two or more wave functions. A vibrating string in the macroscopic universe can be represented by either a single wave or a superposition of several waves (Greenberger et al., 1993).

If and each individually fulfills the Schrödinger equation, then the combined wave function :: also fulfills the Schrödinger equation, where 1 and 2 are complex numbers. The Schrödinger equation is a linear mathematical equation. According to the superposition principle, the combination of many wave functions, which describe the different possible physical states of a system (Nguyen et al., 2013), results in a new wave function that reflects a potential physical state of the system:

$$|\psi\rangle = \sum_i \alpha_i |\psi_i\rangle, \tag{8}$$

where the are complex numbers. The quantity:

$$P = \left|\sum_i \alpha_i |\psi_i\rangle\right|^2, \tag{9}$$

denotes the probability of this superposition. If the states are mutually orthonormal, the probability will be equal to the sum of the individual probabilities:

$$P = \left|\sum_i \alpha_i |\psi_i\rangle\right|^2 = \sum_i |\alpha_i|^2 = P_1 + P_2 + P_3 + \cdots, \quad 10)$$

where Pi = ; Pi is the probability of finding the system in the state (Thomas, 1983).

2.3. OBSERVABLES AND OPERATORS

An observable is a measurable variable that represents a dynamic property. In classical mechanics, the most encountered dynamical variables are position, linear momentum, angular momentum, and energy. What is the mathematical representation of these and other variables in quantum mechanics? The second postulate states that a Hermitian operator is linked to each physical observable. In the previous chapter, it was learned that the position representation of the linear momentum operator is expressed as in one-dimensional space and three-dimensional space (Brodlie & Kisil, 2003).

Typically, any function, that relies on the momentum and position variables, r and p, can be "quantized" or transformed into a function of operators by substituting r $^{\rightarrow}$ and p $^{\rightarrow}$ with their respective operators:

$$f(\vec{r},\vec{p}) \rightarrow F(\hat{\vec{R}},\hat{\vec{P}}) = f(\hat{\vec{R}},-i\hbar\vec{\nabla}), \quad (11)$$

In particular, the operator defining the Hamiltonian:

$$H = \frac{1}{2m}\vec{p}^2 + V(\hat{\vec{r}},t) \quad (12)$$

is given in the position representation by:

$$\hat{H} = -\frac{\hbar^2}{2m}\nabla^2 + V(\hat{\vec{R}},t), \quad (13)$$

where is the Laplacian operator; it is given in Cartesian coordinates by:

Since the momentum operator; is Hermitian, and if the potential is a real function, the Hamiltonian Eqn. (12) is Hermitian, and the eigenvalues of Hermitian operators are real. Hence, the spectrum of the Hamiltonian, which consists of the entire set of its eigenvalues, is real (de Muynck, 1984). This spectrum can be discrete, continuous, or a mixture of both. In the case of bound states, the Hamiltonian has a discrete spectrum of values and a continuous spectrum for unbound states. In general, an operator will have bound or unbound spectra in the same manner that the corresponding classical variable has bound or unbound orbits. As for and they have continuous spectra, since r and p may take a continuum of values (Donnelly & Giddings, 2016).

Postulate 5 states that the total energy E of time-dependent systems is linked to the operator:

$$\hat{H} = i\hbar\frac{\partial}{\partial t}. \quad (14)$$

This can be observed in the following manner. The wave function of a free particle with momentum (; and total energy E is expressed by , where A is a constant. The derivative of with respect to time gives:

$$i\hbar\frac{\partial\psi(\vec{r},t)}{\partial t} = E\psi(\vec{r},t). \quad (15)$$

The next section will examine the eigenfunctions and eigenvalues of the momentum operator . The eigenvalue equation:

$$-i\hbar\vec{\nabla}\psi(\vec{r}) = \vec{p}\psi(\vec{r}) \quad (16)$$

Yields the eigenfunction corresponding to the eigenvalue such that is the probability of finding the particle with a momentum p in the volume element centered about . The solution to the eigenvalue Eqn. (16) is:

$$\psi(\vec{r}) = Ae^{i\vec{p}.\vec{r}/\hbar},$$

where; A is a normalization constant (Doplicher et al., 1971). Since ; is the eigenvalue of the operator , the eigenfunction Eqn. (17) reduces to hence the eigenvalue Eqn. (16) becomes:

$$\hat{\vec{P}}\psi(\vec{r}) = \hbar\vec{k}\psi(\vec{r}). \quad (17)$$

2.4. MEASUREMENT IN QUANTUM MECHANICS

Quantum theory just pertains to the outcomes of measurement and does not provide any information regarding events that may occur in physical reality beyond the scope of measurement. The focus is primarily on measurement (Lockhart & Misra, 1986).

2.4.1. How Measurements Disturb Systems

In the realm of classical physics, it is feasible to conduct measurements on a system without causing significant disruption to it. However, in the realm of quantum physics, the act of measurement causes a considerable disturbance to the system. During the process of measuring classical systems, a disturbance does indeed exist, but its magnitude is sufficiently minor to be disregarded. However, in atomic and subatomic systems, the process of measurement causes noticeable or substantial disruptions (Kalgotra et al., 2019).

For example, suppose an experiment that analyzes the precise location of an electron in a hydrogen atom. To do this, it is necessary to subject the electron to electromagnetic radiation (photons) by bombardment. In order to achieve precise positioning, it is necessary for the wavelength of the radiation to be sufficiently small. Given that the size of the electronic orbit is approximately m, it is necessary to employ radiation with a wavelength smaller than m. In other words, it is necessary to subject the electron to a bombardment of photons with energy greater than (Tréangle et al., 2019):

$$h\nu = h\frac{c}{\lambda} = h\frac{3\times10^{8}}{10^{-10}} \sim 10^{4}eV. \tag{18}$$

When these photons collide with the electron, they not only disturb it but also dislodge it from its orbit entirely. It is important to note that the ionization energy of the hydrogen atom is around 13.5 electron volts (eV). Hence, the simple act of determining the electron's position significantly disrupts it (Busch et al., 2014).

Now, let's delve into the fundamental idea of measurement in quantum physics. Measurement often induces a modification in the state of the system. Theoretically, one can use an operator to symbolize the measuring instrument. This allows the system to be in one of the eigenstates of the operator once the measurement is performed (Iqbal et al., 2005). Examine a system that is in a state . Prior to measuring an observable A, the state can be expressed as a linear combination of eigenstates of the corresponding operator

$$|\psi\rangle = \sum_n |\psi_n\rangle\langle\psi_n|\psi\rangle = \sum_n a_n|\psi_n\rangle. \tag{19}$$

Postulate 4 asserts that while measuring property A, the system's state changes from to one of the eigenstates of the operator . The measured result corresponds to the eigenvalue a_n. The sole deviation from this principle occurs when the system is already in one of the eigenstates of the observable being assessed (Cannon et al., 1995). For example, if the system is in the eigenstate , a measurement of the observable A will produce the value with a probability of 1, and it will not alter the state

Prior to a measurement, one can only predict with probability the eigenstate in which a system will end up after the measurement among the several states . Postulate 4 asserts that the likelihood of observing the system in a specific nondegenerate eigenstate | is determined by:

$$P_n = \frac{|\langle\psi_n|\psi\rangle|^2}{\langle\psi|\psi\rangle}. \tag{20}$$

It is important to understand that the wave function does not provide specific predictions for individual observations. Rather, it establishes the probability distribution, denoted as , for measurements conducted on multiple matching systems in the same state (Bills & Evett, 1959).

In conclusion, quantum mechanics is the branch of mechanics that applies to things when measurements unavoidably disrupt the system's state. From a quantum mechanical perspective, it is crucial to consider the impact of the measuring equipment on the system, as it cannot be disregarded. Typically, specific measurements cannot be carried out without significantly affecting other characteristics of the quantum system. Ultimately, the fundamental aspect of quantum mechanics lies in the impact of equipment interference on the system (Dong et al., 2010).

2.4.2. Expectation Values

The expected value of the operator with regard to the state is mathematically defined as the average value of :

$$\langle\hat{A}\rangle = \frac{\langle\psi|\hat{A}|\psi\rangle}{\langle\psi|\psi\rangle}. \tag{21}$$

For example, the energy of a system is determined by the average value of the Hamiltonian:

$$E = \langle \hat{H} \rangle = \langle \psi | \hat{H} | \psi \rangle / \langle \psi | \psi \rangle. \quad (22)$$

The expected value indicates the mean outcome of calculating on the state . In order to observe this, one can utilize the entire collection of eigenvectors of as a basis (meaning is diagonal in). Consequently, one can rephrase in the following manner (Bartlett & Noga, 1988):

$$\langle \hat{A} \rangle = \frac{1}{\langle \psi | \psi \rangle} \sum_{nm} \langle \psi | \psi_m \rangle \langle \psi_m | \hat{A} | \psi_n \rangle \langle \psi_n | \psi \rangle = \sum_n a_n \frac{|\langle \psi_n | \psi \rangle|^2}{\langle \psi | \psi \rangle}, \quad (23)$$

The associated location is at the inner product of the state vectors . The expression represents the probability of observing the value while measuring the observable A. Therefore, one can interpret as the average result obtained from a series of measurements of A:

$$\boxed{\langle \hat{A} \rangle = \sum_n a_n \frac{|\langle \psi_n | \psi \rangle|^2}{\langle \psi | \psi \rangle} = \sum_n a_n P_n.} \quad (24)$$

The anticipated outcome of an observable is calculated by summing all allowable eigenvalues, where each eigenvalue is multiplied by its corresponding probability The Eqn. (25), which holds true for discrete spectra (Klahn & Morgan, 1984), can be generalized to encompass a continuous probability distribution P(a) in the following manner:

$$\langle \hat{A} \rangle = \frac{\int_{-\infty}^{+\infty} a \, |\psi(a)|^2 da}{\int_{-\infty}^{+\infty} |\psi(a)|^2 da} = \int_{-\infty}^{+\infty} a \, dP(a). \quad (25)$$

The physical method for obtaining the expected value of an observable is as follows: Prepare a vast quantity of identical systems, each in an identical state . Next, the identical systems are subjected to measurements of the observable A, resulting in outcomes , ...; The probabilities of these outcomes occurring are denoted as , The expectation value of Aˆ with regard to the state is the average value obtained from multiple repeated measurements (Barnett et al., 1991).

It is important to understand that in classical physics, measuring the same observable on several identically prepared systems should yield the same result. However, in certain cases, different results can be obtained while evaluating the same observable on multiple equally prepared systems, which goes against classical physics. In the realm of quantum physics, the capacity to make predictions is limited to determining the probability of attaining a specific value for an observable (Livadiotis, 2012).

2.4.3. Complete Sets of Commuting Operators (CSCO)

Two observables A and B are said to be compatible when their corresponding operators commute; observables corresponding to non-commuting operators are said to be non-compatible (Cerrato & Noce, 2019).

The next section will examine the process of quantifying two variables, A and B, on a specific system. Due to the inherent disturbance caused by the act of measurement, the outcome of measuring A and B is contingent upon the sequence in which they are performed. When A is measured before B, the resulting outcomes tend to differ from those obtained when B is measured before A (Chen et al., 1985).

If and do not commute and the system is in an eigenstate of , then a measurement of A will definitely produce a value. Subsequently, upon measuring B, the system will be found in one of the eigenstates of B. If another measurement of A is conducted, a value will be obtained that will differ from the previous measurement of A. What is the meaning or significance of this newly assigned value? It is not feasible to provide a definitive response to this issue; only a probabilistic outcome may be determined. To accomplish this, it is necessary to express the eigenstates of B in relation to those of A, hence offering a probabilistic determination of the measurement value of A. If and do not commute, their simultaneous measurement is not possible; the sequence of their measurement is significant (Tarasov, 2008).

What is the result when both A and B transfer? One can demonstrate that the outcomes of their measures will be independent of the sequence in which they are conducted. Prior to presenting this, it is important to mention a valuable theorem (Lozada et al., 1992).

2.4.4. Measurement and the Uncertainty Relations

The uncertainty requirement for measuring both observables A and B is expressed as follows:

$$\Delta A \Delta B \geq \frac{1}{2}|\langle[\hat{A},\ \hat{B}]\rangle|, \tag{26}$$

This is demonstrated by using the combined measurement of the position and momentum observables. Due to the incompatibility of both observables, it is not possible to measure them simultaneously with infinite accuracy. In other words, since there is no state that can be an eigenstate of both and at the same time (Busch & Stevens, 2015). In the context of the position and momentum operators, Eqn. (26) provides the following result:

$$\Delta x\, \Delta p \geq \frac{\hbar}{2}. \tag{27}$$

This condition demonstrates that it is impossible to estimate a tiny system's position and momentum simultaneously with infinite accuracy. The uncertainty in the measurement of momentum associated with a location uncertainty Δx cannot be less than This is a result of the interference that occurs between the two measurements. By initially measuring the position, the system is disturbed by altering its state to an eigenstate of the position operator. Subsequently, while measuring the momentum, the system is forced into an eigenstate of the momentum operator. An intriguing use of the uncertainty relation

Eqn. (27) pertains to the orbital angular momentum of a particle (Breitenberger, 1985). Given that the components of the commutator one can derive the following result:

$$\Delta L_x \Delta L_y \geq \frac{1}{2}\hbar|\langle \hat{L}_z \rangle|. \tag{28}$$

One can obtain the other two inequalities by means of a cyclic permutation of x, y, and z. If NLˆ will have sharp values simultaneously. This occurs when the particle is in an s state. In fact, when a particle is in an s state, ; hence all the components of orbital angular momentum will have sharp values simultaneously (Barchielli et al., 2018).

Remember

In quantum mechanics, measurement results in the collapse of the wave function, which modifies the system's state.

2.5. MATTER WAVES

A charged pith ball's electric field extends over space rather than being entirely confined. Consider the charged particle and the field to be two facets of the same thing, or inseparable. The tests conducted in Chapter 1, specifically those pertaining to the scattering of electrons and atoms, demonstrate that it is impossible to fully confine matter on an atomic level. The concept of incompletely localized charges is more familiar than that of unlocalized matter, yet both phenomena are equally enigmatic. The fundamental experiments can be elucidated quantitatively by postulating that each unit of matter is accompanied by a novel form of field denoted by the symbol 'Ψ,' which possesses wave-like properties (Brazhnyi et al., 2004).

The wavelength of these waves in free space is determined by the de Broglie equation ⊠ = h/mv, which is derived from the theory. The behavior of particles is closely linked to the waves they are associated with. To accurately forecast the behavior of particles, it is necessary to understand the current state of the Ψ waves. The fundamental principles of quantum physics provide guidelines for computing the total wave function Ψ(x,y,z,r) of any given system. Furthermore, they provide instructions on how to compute the anticipated value of measurable quantities using the Ψ function. The theory is unable to forecast the specific behavior of individual systems, but it can only provide insights into the average behavior of a substantial number of systems (Kozuma et al., 1999). The Davisson-Germer experiment demonstrates the limitation of predicting the average behavior of multiple systems. By replacing the "Faraday cage" with a counter device that detects individual electrons, it is observed that electrons are detected at various angles. However, the overall count reveals a distinctive interference pattern. Regardless, the theory of matter waves solely focuses on making predictions about the average behavior of several systems. There appears to be no alternative method to accurately interpret experimental observations (Müller et al., 2010). Like the E and B fields of electricity, the '-function is not itself directly observable. It is a tool for calculation. Since it gives results that are in agreement with the experiment it has a certain degree of reality. As shall be seen, the * function is the only contact the macroscopic world has with the microscopic world. One might say that * is the "window to the world of the atom." what is not revealed by the wave function using the methods of the theory cannot be found out, and calculations using the theory always involve both . Quantum mechanics was first formulated in terms of matrix algebra by Heisenberg. An equivalent form, independently discovered by Schrodinger, is known as wave mechanics (Lücke et al., 2011).

The Ψ-function is explicitly present in Schrodinger's formulation, making it the most accessible form of the theory to comprehend. The phrases wave mechanics and quantum mechanics have become almost interchangeable due to their fundamental similarities (Anker et al., 2005).

ACTIVITY 2.1

Objective: Understand the concepts of quantum states and probability densities through practical application.

Materials Needed:

- Graph paper or graphing software
- Access to information on a simple wave function
- Calculator (optional)

Instructions:

a. Wave Function Representation:
   - Choose a simple wave function and define its characteristics. Write a brief description of this wave function on paper.

b. Graphing the Wave Function:
   - Plot the wave function on graph paper or using graphing software over a specified range (e.g., from –3 to 3).

c. Calculating Probability Density:
   - Based on your wave function, determine the probability density and create a visual representation of this density on the same graph.

d. Discussion Questions:
   - How does the shape of the wave function compare to the probability density?
   - What does the probability density indicate about the likelihood of finding a particle at various positions?

e. Reflection:
   - Write a brief summary of your findings, discussing the implications of the wave function and probability density in understanding quantum systems.

SUMMARY

- The fundamental concepts of quantum mechanics are presented in the chapter on the Postulates of Quantum Mechanics. The study commences by establishing the fundamental principles, highlighting the distinctive characteristics of quantum phenomena in contrast to classical physics. The state of a system is described by a wave function or state vector, which contains all the information about the system and enables the computation of probability densities, indicating the chances of locating a particle in particular regions of space.
- Important principles in this context are the superposition principle, which permits systems to exist in numerous states at the same time until they are observed, and the significance of observables, which are represented by operators. The measurement processes play a critical role in quantum mechanics since they cause the wave function to collapse into an eigenstate, hence affecting the state of the system and subsequent measurements.
- In addition, the chapter explores expectation values, the importance of complete sets of commuting operators (CSCO), and Heisenberg's uncertainty principle, which emphasizes inherent constraints on the simultaneous measurement of specific pairs of observables. The notion of matter waves is also presented, demonstrating the underlying wave-particle duality seen in quantum systems.

REVIEW QUESTIONS

1. What are the basic postulates of quantum mechanics, and how do they differ from classical mechanics?
2. Explain the significance of the wave function in representing the state of a quantum system. What information does it contain?
3. Describe the superposition principle and provide an example of how it manifests in quantum systems.
4. How do observables relate to operators in quantum mechanics? What role do they play in measurements?
5. Discuss the process of measurement in quantum mechanics and its effect on the state of a system. What is meant by wave function collapse?
6. What is the meaning of expectation values, and how are they calculated in quantum mechanics?
7. Define a complete set of commuting operators (CSCO) and explain its importance in the context of simultaneous measurements.
8. Summarize Heisenberg's uncertainty principle. What implications does it have for the measurement of complementary observables?

MULTIPLE CHOICE QUESTIONS

1. **What does the wave function represent in quantum mechanics?**
   a. The exact position of a particle
   b. The potential energy of a system
   c. The state of a quantum system
   d. The mass of a particle
2. **What principle states that a quantum system can exist in multiple states simultaneously?**
   a. Complementarity
   b. Superposition
   c. Entanglement
   d. Uncertainty
3. **In quantum mechanics, what are observables represented by?**
   a. Scalars
   b. Vectors
   c. Matrices (Operators)
   d. Functions
4. **What happens to the wave function upon measurement of an observable?**
   a. It remains unchanged
   b. It becomes a superposition
   c. It evolves continuously
   d. It collapses to an eigenstate
5. **Heisenberg's uncertainty principle indicates a limitation on the simultaneous measurement of which of the following pairs?**
   a. Energy and mass
   b. Position and momentum
   c. Spin and charge
   d. Velocity and acceleration
6. **What is meant by the term "complete set of commuting operators" (CSCO)?**
   a. A set of operators that share the same eigenstates
   b. A set of operators that cannot be measured together
   c. A set of operators that have different dimensions
   d. A set of operators that are always zero

Answers to Multiple Questions

1. (c); 2. (b); 3. (c); 4. (d); 5. (b); 6. (a)

REFERENCES

1. Anker, T., Albiez, M., Gati, R., Hunsmann, S., Eiermann, B., Trombettoni, A., & Oberthaler, M. K. (2005). Nonlinear self-trapping of matter waves in periodic potentials. Physical Review Letters, 94(2), 020403.
2. Barchielli, A., Gregoratti, M., & Toigo, A. (2018). Measurement uncertainty relations for discrete observables: Relative entropy formulation. Communications in Mathematical Physics, 357(3), 1253–1304.
3. Barnett, R. N., Reynolds, P. J., & Lester Jr, W. A. (1991). Monte Carlo algorithms for expectation values of coordinate operators. Journal of Computational Physics, 96(2), 258–276.
4. Bartlett, R. J., & Noga, J. (1988). The expectation value coupled-cluster method and analytical energy derivatives. Chemical Physics Letters, 150(1–2), 29–36.
5. Bills, D. G., & Evett, A. A. (1959). Glass, a disturbing factor in physical electronics measurements. Journal of Applied Physics, 30(4), 564–567.
6. Brazhnyi, V. A., & Konotop, V. V. (2004). Theory of nonlinear matter waves in optical lattices. Modern Physics Letters B, 18(14), 627–651.
7. Breitenberger, E. (1985). Uncertainty measures and uncertainty relations for angle observables. Foundations of Physics, 15(3), 353–364.
8. Brodlie, A., & Kisil, V. V. (2003). Observables and states in p-mechanics. Advances in Mathematics Research, 5, 101–136.
9. Busch, P., & Stevens, N. (2015). Direct tests of measurement uncertainty relations: What it takes. Physical Review Letters, 114(7), 070402.
10. Busch, P., Lahti, P., & Werner, R. F. (2014). Colloquium: Quantum root-mean-square error and measurement uncertainty relations. Reviews of Modern Physics, 86(4), 1261–1281.
11. Cai, B., Hennessy, S., Lo Re III, V., & Small, D. S. (2015). Epidemiologic research using probabilistic outcome definitions. Pharmacoepidemiology and Drug Safety, 24(1), 19–26.
12. Cannon, P. S., Davies, N. C., Angling, M. J., Jodalen, V., Moreland, K. W., & Lundborg, B. (1995). Initial results from DAMSON—A system to measure multipath, Doppler spread and Doppler shift on disturbed HF channels. In 1995 Ninth International Conference on Antennas and Propagation, ICAP'95 (Conf. Publ. No. 407, Vol. 2, pp. 104–108).
13. Carcassi, G., Maccone, L., & Aidala, C. A. (2021). Four postulates of quantum mechanics are three. Physical Review Letters, 126(11), 110402.
14. Cerrato, N., & Noce, C. (2019). Complete set of commuting observables for a two-site Hubbard model. European Journal of Physics, 40(5), 055403.
15. Chen, J. Q., Gao, M. J., & Ma, G. Q. (1985). The representation group and its application to space groups. Reviews of Modern Physics, 57(1), 211.
16. de Muynck, W. M. (1984). A quantum mechanical theory of local observables and local operations. Foundations of Physics, 14(3), 199–253.

17. Debs, A. S., & Larson, R. E. (1970). A dynamic estimator for tracking the state of a power system. IEEE Transactions on Power Apparatus and Systems, (7), 1670–1678.

18. Dong, Y., Hu, Z., Uchimura, K., & Murayama, N. (2010). Driver inattention monitoring system for intelligent vehicles: A review. IEEE Transactions on Intelligent Transportation Systems, 12(2), 596–614.

19. Donnelly, W., & Giddings, S. B. (2016). Observables, gravitational dressing, and obstructions to locality and subsystems. Physical Review D, 94(10), 104038.

20. Doplicher, S., Haag, R., & Roberts, J. E. (1971). Local observables and particle statistics I. Communications in Mathematical Physics, 23, 199–230.

21. Faraggi, A. E., & Matone, M. (2000). The equivalence postulate of quantum mechanics. International Journal of Modern Physics A, 15(13), 1869–2017.

22. Gillespie, D. T. (1976). A general method for numerically simulating the stochastic time evolution of coupled chemical reactions. Journal of Computational Physics, 22(4), 403–434.

23. Greenberger, D. M., Horne, M. A., & Zeilinger, A. (1993). Multiparticle interferometry and the superposition principle. Physics Today, 46(8), 22–29.

24. Gudder, S. P., & Mullikin, H. C. (1973). Measure theoretic convergences of observables and operators. Journal of Mathematical Physics, 14(2), 234–242.

25. Hardy, J., Pomeau, Y., & De Pazzis, O. (1973). Time evolution of a two-dimensional model system. I. Invariant states and time correlation functions. Journal of Mathematical Physics, 14(12), 1746–1759.

26. Haworth, D. C. (2010). Progress in probability density function methods for turbulent reacting flows. Progress in Energy and Combustion Science, 36(2), 168–259.

27. Iqbal, S., Gualini, M. M. S., & Rashid, K. (2005). Stray noises and illuminations disturbing the performance of position-sensitive devices. In Optical Modeling and Performance Predictions II (Vol. 5867, pp. 36–45).

28. Kalgotra, P., Sharda, R., & McHaney, R. (2019). Don't disturb me! Understanding the impact of interruptions on knowledge work: An exploratory neuroimaging study. Information Systems Frontiers, 21, 1019–1030.

29. Klahn, B., & Morgan III, J. D. (1984). Rates of convergence of variational calculations and of expectation values. The Journal of Chemical Physics, 81(1), 410–433.

30. Klyshko, D. N. (1998). Basic quantum mechanical concepts from the operational viewpoint. Physics-Uspekhi, 41(9), 885.

31. Koliander, G., El-Laham, Y., Djuri⊠, P. M., & Hlawatsch, F. (2022). Fusion of probability density functions. Proceedings of the IEEE, 110(4), 404–453.

32. Kozuma, M., Suzuki, Y., Torii, Y., Sugiura, T., Kuga, T., Hagley, E. W., & Deng, L. (1999). Phase-coherent amplification of matter waves. Science, 286(5448), 2309–2312.

33. Livadiotis, G. (2012). Expectation values and variance based on L p-norms. Entropy, 14(12), 2375–2396.

34. Lockhart, C. M., & Misra, B. (1986). Irreversibility and measurement in quantum mechanics. Physica A: Statistical Mechanics and its Applications, 136(1), 47–76.

35. Lozada, A., Sarrazin, G. D., & Torres, P. L. (1992). Complete sets of commuting observables and irreducible sets of observables. Helvetica Physica Acta, 65(7), 887–897.
36. Lücke, B., Scherer, M., Kruse, J., Pezzé, L., Deuretzbacher, F., Hyllus, P., & Klempt, C. (2011). Twin matter waves for interferometry beyond the classical limit. Science, 334(6057), 773–776.
37. Marshman, E., & Singh, C. (2019). Validation and administration of a conceptual survey on the formalism and postulates of quantum mechanics. Physical Review Physics Education Research, 15(2), 020128.
38. Masanes, L., Galley, T. D., & Müller, M. P. (2019). The measurement postulates of quantum mechanics are operationally redundant. Nature Communications, 10(1), 1361.
39. Masanes, L., Müller, M. P., Augusiak, R., & Pérez-García, D. (2013). Existence of an information unit as a postulate of quantum theory. Proceedings of the National Academy of Sciences, 110(41), 16373–16377.
40. Mielnik, B. (1974). Generalized quantum mechanics. Communications in Mathematical Physics, 37, 221–256.
41. Müller, H., Peters, A., & Chu, S. (2010). A precision measurement of the gravitational redshift by the interference of matter waves. Nature, 463(7283), 926–929.
42. Nguyen, Q. T., Di Benedetto, H., Sauzéat, C., & Tapsoba, N. (2013). Time temperature superposition principal validation for bituminous mixes in the linear and nonlinear domains. Journal of Materials in Civil Engineering, 25(9), 1181–1188.
43. Omnes, R. (1988). Logical reformulation of quantum mechanics. I. Foundations. Journal of Statistical Physics, 53, 893–932.
44. Paris, M. G. (2012). The modern tools of quantum mechanics: A tutorial on quantum states, measurements, and operations. The European Physical Journal Special Topics, 203(1), 61–86.
45. Primadianto, A., & Lu, C. N. (2016). A review on distribution system state estimation. IEEE Transactions on Power Systems, 32(5), 3875–3883.
46. Schmiedmayer, J., Chapman, M. S., Ekstrom, C. R., Hammond, T. D., Wehinger, S., & Pritchard, D. E. (1995). Index of refraction of various gases for sodium matter waves. Physical Review Letters, 74(7), 1043.
47. Tarasov, V. E. (2008). Spectral representation of observable. Monograph Series on Nonlinear Science and Complexity, 7, 181–209.
48. Thomas, A. S. (1983). The control of boundary-layer transition using a wave-superposition principle. Journal of Fluid Mechanics, 137, 233–250.
49. Tréangle, C., Farza, M., & M'Saad, M. (2019). Observer design for a class of disturbed nonlinear systems with time-varying delayed outputs using mixed time-continuous and sampled measurements. Automatica, 107, 231–240.
50. Van den Berg, M., & Iversen, M. (2013). On the minimization of Dirichlet eigenvalues of the Laplace operator. Journal of Geometric Analysis, 23(2), 660–676.
51. Wu, S., Wang, Y. J., Diot, Q., & Prentiss, M. (2005). Splitting matter waves using an optimized standing-wave light-pulse sequence. Physical Review A: Atomic, Molecular, and Optical Physics, 71(4), 043602.

CHAPTER 3

Schrodinger Equation

LEARNING OBJECTIVES

After studying this chapter, you will be able to:

1. Understand the concept and properties of waves.
2. Comprehend the difference between dispersive and non-dispersive waves
3. Understand a free particle Schrodinger equation
4. Comprehend the Schrodinger equation of a particle having potential energy

KEY TERMS FROM THIS CHAPTER

Dispersive Waves
Linear Superpositions
Schrodinger Equation
Particle
Sinusoidal Waves

UNIT INTRODUCTION

The formulation of a wave equation that can describe the stealthy, wave-like behavior of a quantum particle is the initial stage in the process of developing a theory of nonrelativistic quantum mechanics that is consistent with logic (Feit et al., 1982). The Schrodinger equation is the name given to this equation. When it comes to quantum mechanics, the Schrodinger equation serves a function that is comparable to that of Newton's Laws within classical fundamental physics. Movement is described by both (Schleich et al., 2013).

3.1. WAVES

To prepare for the Schrodinger equation, this section will first go over the ways in which mathematics may be utilized to describe waves of varying sizes and formations (Horton, 1999).

3.1.1. Sinusoidal Waves

A sinusoidal traveling wave has a specific wavelength λ and period τ, which may be expressed as a definite wave number, $k = 2p\pi/\lambda$, and an angular frequency, $\omega = 2\pi v/\lambda$ (Kosevich, 1993). One possible mathematical function that may be used to express such a wave is:

$$\Psi(x,t) = A\cos(kx - \omega t)$$

where A is a fixed constant. At every point x, the function ψ (x, t) begins to oscillate with an amplitude of A and a time of $2\pi/\omega$. Oscillating with an amplitude of A and a wavelength of $2\pi/k$, the function ψ (x, t) is observed to oscillate at each time t. Furthermore (, it is worth noting that these oscillations propagate in a manner wherein the velocity of x increases with a velocity of ω/k. For instance, the highest point of ψ (x, t) when $kx - \omega t = 0$ is located at $x = \omega t/k$, while the lowest point when $kx - \omega t = \pi$ is located at $x = \lambda/2 + \omega t/k$.

A sinusoidal traveling wave is described by the function sin $(kx - \omega t)$, which is like the function cos $(kx - \omega t)$ in that it also describes the wave equation:

$$\sin(kx - \omega t) = \cos(kx - \omega t - \pi/2),$$

The sin $(kx - \omega t)$ as well as cos $(kx - \omega t)$ waves exhibit a phase difference of $\pi/2$, meaning that their undulations and oscillations are not synchronized. The linear superposition represents the most general form of such waves:

$$\Psi(x,t) = A\cos(kx - \omega t) + B\sin(kx - \omega t),$$

where A and B are variables that can take on any value.

In the field of classical physics, sinusoidal traveling waves are typically described by complex exponential equations of the type (Greenhow & White, 1997). In quantum physics, on the other hand, they are always represented in this manner:

$$\Psi(x,t) = A\,e^{i(kx - \omega t)}.$$

The utilization of complex exponentials to depict waves in classical physics is purely a mathematical accessibility (Kuntz & Blackstock,1987). As an illustration, a sound wave pressure can be represented by the actual function Acos $(kx - \omega t)$. It may be fragment of Aei$(kx - \omega t)$ as:

$$e^{i(kx-\omega t)} = \cos(kx - \omega t) + i\sin(kx - \omega t).$$

Did you know?

Erwin Schrödinger developed his famous equation in 1926, earning him a Nobel Prize in Physics in 1933.

Using the real component of a complex exponential, it is possible to express a real sinusoidal wave in classical physics (Okanuma, 2006). This is possible because of several factors. The application of complex numbers is not something that can be considered a matter of choice in the field of quantum physics. Upon closer inspection, it becomes apparent that any complex exponential can provide an appropriate representation of a de Broglie wave (Nayfeh, 1974).

3.1.2. Linear Superpositions of Sinusoidal Waves

Standing waves can be created by combining two sinusoidal waves that are traveling in opposite directions. For instance, the linear superposition:

$$A\cos(kx - \omega t) + A\cos(kx + \omega t)$$

Produces the wave *2Acoskx.* This wave exhibits oscillations with a period of $2\pi/\omega$ and undulations with a wavelength of $2\pi/k$. However, it does not propagate, but rather remains stationary and oscillates without moving. It can be described as a non-propagating wave (Budiyono & Umeno, 2009).

An instance of this may be seen in the mathematical expression of a wave packet (Nimmo & Schief, 1997), which is created by combining many sinusoidal waves with a consistent amplitude A as well as wave numbers ranging from k-Δk to k + Δk:

$$\Psi(x,t) = \int_{k-\Delta k}^{k+\Delta k} A\cos(k'x - \omega' t)\,dk'.$$

In the event that the value of k is positive, the wave packet will transfer in the direction of positive x.

One can calculate the integral in order to obtain the initial form of the wave packet, which is the shape at the time when time equals zero (Madsen & Schaeffer, 2010):

$$\Psi(x,0) = \int_{k-\Delta k}^{k+\Delta k} A\cos k'x\,dk'.$$

$$\Psi(x,0) = S(x)\cos kx, \qquad \text{where} \quad S(x) = 2A\Delta k\frac{\sin(\Delta kx)}{(\Delta kx)}.$$

If the difference (Δk) is significantly smaller than the wave number (k), a sinusoidal wave (coskx) that exhibits rapid variations is observed. A function S(x) that gradually varies, which achieves its highest value at x = 0 and has zeros when x is a multiple of an integer of π/Δk, is responsible for regulating the amplitude of this wave. A wave packet is produced as a result, and its effective length is approximately 2 times the square of the kth derivative (Donelan et al., 1992). There are three wave packets depicted in Figure 3.1, each of which is distinguished by a different value of Δk. It has been noticed that the wave packets acquire a larger length as the range associated with wave numbers decreases. Furthermore, when the value of Δk approaches 0, these wave packets undergo a transformation into "monochromatic" waves that have an indefinite duration. Other types of wave packets exhibit behavior that is comparable to this (Franaszczuk & Blinowska, 1985). The link between the wave number and the angular frequency is the primary factor that determines the speed at which a wave packet moves as well as the potential for the packet's shape to change while it is propagating (Babanin & Rogers, 2014).

Did you know?

According to quantum mechanics, particles can exist in multiple states at once. This principle is key to the operation of quantum computers, which can perform many calculations simultaneously.

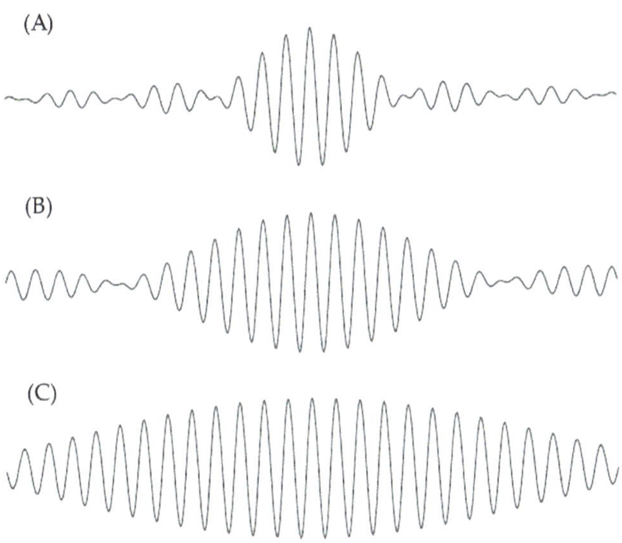

Figure 3.1. *Illustration of the initial shapes of the wave packets given by a linear superposition of sinusoidal waves with constant amplitude A and wave numbers in the range k – Δk to k + Δk.*

Note: The three diagrams show how the length of a wave packet increases as the range of wave numbers Δk decreases. The value of AΔk is constant, but Δk equals k/8 in diagram (A); Δk equals k/16 in diagram (B); and Δk equals k/32 in diagram (C). In general, the length of a wave packet is inversely proportional to Δk and becomes infinite in extent as Δk⊠ 0.

Source: A.C. Phillips, Creative Commons License.

3.1.3. Dispersive and Non-Dispersive Waves

A non-dispersive wave is best shown by the electromagnetic radiation that occurs in a vacuum. This is the most well-known scenario. A wave that is not dispersive is defined by a dispersion relation that is represented as ω = ck, where c is a constant (Sheremet & Guza, 1999). Therefore, it can be concluded that the velocity of a sinusoidal wave, which is represented by the equation ω/k = c, remains constant regardless of the wave number k. In the process of travel, a wave packet, which is produced by merging sinusoidal waves in a linear superposition, can keep its shape because the velocities of each sinusoidal component are identical (Buchleitner et al., 2002).

The form of waves traveling in three dimensions is as follows:

$$\nabla^2\Psi - \frac{1}{c^2}\frac{\partial^2\Psi}{\partial t^2} = 0, \quad \text{where} \quad \nabla^2 = \frac{\partial^2}{\partial x^2} + \frac{\partial^2}{\partial y^2} + \frac{\partial^2}{\partial z^2},$$

For waves propagating in a single dimension, such as the x direction, their shape can be described as follows:

$$\frac{\partial^2\Psi}{\partial x^2} - \frac{1}{c^2}\frac{\partial^2\Psi}{\partial t^2} = 0.$$

As an illustration, the sinusoidal waves,

$$A\cos(kx - \omega t), \qquad A\sin(kx - \omega t), \qquad \text{or } A\,e^{i(kx-\omega t)},$$

The solutions presented are ω2 = c2k, which can be demonstrated by directly substituting into the equation above. Since every parameter in the classical wave equation remains linear with respect to ψ, it follows that a solution can be obtained by combining sinusoidal waves through linear superposition. For instance, a superposition refers to a solution that explains a wave packet that propagates without any alteration in its shape (Didenkulova & Pelinovsky, 2009).

Nevertheless, most waves found in both conventional and quantum physics exhibit the property of dispersion. Therefore, a group of waves that spread out will typically alter its form while it moves forward (Givoli & Neta, 2003). The term used to describe this velocity is known as the group velocity, which is determined by:

$$\upsilon_{\text{group}} = \frac{d\omega}{dk},$$

The movement of a concentrated disturbance that is the consequence of the constructive combination of many sinusoidal waves is characterized by the group velocity. This occurs when the waves are in phase, meaning that:

$$k_1 x - \omega_1 t = k_2 x - \omega_2 t.$$

By manipulating this equation, one can determine that the location of this juncture of constructive interference is expressed as:

$$x = \left(\frac{\omega_1 - \omega_2}{k_1 - k_2} \right) t.$$

However, it is possible for several sinusoidal waves to combine and create a specific area of constructive interference that moves at a velocity determined by the equations (Zhang et al., 2021).

To demonstrate the derivation of group velocity, examine the case of water waves with large wavelengths that adhere to the dispersion relation:

$$\omega = \sqrt{gk},$$

The phase velocity refers to the velocity that produces a sinusoidal water wave (Nambu, 1966), is:

$$\upsilon_{phase} = \frac{\omega}{k} = \sqrt{\frac{g}{k}},$$

The velocity of a packet of water waves, which consists of wave numbers around k within a narrow range, is:

$$\upsilon_{group} = \frac{d\omega}{dk} = \frac{1}{2}\sqrt{\frac{g}{k}},$$

Put simply, the sinusoidal waves that make up the packet travel at twice the velocity of the place where the interference of the waves creates the greatest disruption. Nevertheless, the form of the disruption will alter as it travels; typically, it will tend to disperse (Simonis et al., 1984).

Did you know?

According to quantum mechanics, particles can exist in multiple states at once. This principle is key to the operation of quantum computers, which can perform many calculations simultaneously.

3.2. PARTICLE WAVE EQUATIONS

To provide an explanation for wave-like occurrences, the wave equations are derived from the fundamental laws of physics within the sphere of classical physics (Chraplyvy, 1953). For example, the classical wave equation, which governs the behavior of electromagnetic waves in empty space, can be derived by using Maxwell's principles of electromagnetism. This equation takes into account the behavior of electromagnetic waves (Bhabha, 1945).

3.2.1. Wave Equation for a Free Particle

The de Broglie wavelength associated with a particle with momentum p is given by the equation $\lambda = h/p$. Therefore, a de Broglie wave characterized by a wave number $k = 2\pi/\lambda$ represents:

$$p = \hbar k, \ \ where\ \hbar = \frac{h}{2\pi}$$

In order to further develop this concept (Winfree, 1990), make the assumption that a de Broglie wave packet, characterized by a range of wave numbers from $k - \Delta k$ to $k + \Delta k$. It represents a particle with an indeterminate momentum:

$$\Delta p \approx \hbar \Delta k.$$

Additionally, presume that the magnitude of this wave packet corresponds to Δx, which represents the uncertainty in the particle's position (Dagan & Bush, 2020). Referring to the equations mentioned above:

$$\Delta x \approx \frac{2\pi}{\Delta k}.$$

Or,

$$\Delta x \Delta p \approx h,$$

Nevertheless, it is important to acknowledge that a de Broglie wave undergoes a transformation when subjected to a measurement. If an accurate measurement of the particle's position is taken, the resulting wave packet must be extremely narrow, consisting of a combination of sinusoidal waves with a broad spectrum of wavelengths and vice versa. This suggests that the wave packet is a delicate entity that undergoes a transformation when subjected to a measurement (Belavkin & Staszewski, 1992). The mechanism behind this phenomenon remains unknown (Roberts & Buchsbaum, 1964).

In other words:

$$\frac{d\omega}{dk} = \frac{\hbar k}{m}.$$

The integration of this equation yields the dispersion relationship for the de Broglie waves that describe a quantum particle with mass m and free motion (Misra, 2012).:

$$\omega = \frac{\hbar k^2}{2m}.$$

By setting the constant for integration to zero, eliminate any observable effects it may have in non-relativistic quantum physics, resulting in the obtained connection.

The objective is to identify a wave equation that possesses sinusoidal solutions that adhere to this dispersion relation (Guerrero et al., 2011). The shape of a free particle traveling in one dimension is given by:

$$i\hbar\frac{\partial \Psi}{\partial t} = -\frac{\hbar^2}{2m}\frac{\partial^2 \Psi}{\partial x^2}.$$

Or,

$$\Psi(x, t) = A\, e^{i(kx-\omega t)}$$

Gives: and the RHS yields;

$$i\hbar\frac{\partial \Psi}{\partial t} = i\hbar(-i\omega)A\, e^{i(kx-\omega t)} = \hbar\omega A\, e^{i(kx-\omega t)},$$

$$-\frac{\hbar^2}{2m}\frac{\partial^2 \Psi}{\partial x^2} = \frac{\hbar^2 k^2}{2m} A\, e^{i(kx-\omega t)},$$

Given that the sinusoidal solution represents a wave with wave number k as well as angular velocity ω, one may presume that it depicts a particle moving freely in the direction of x with a well-defined momentum p = hk with energy E = p2/2m = hω.

The solutions to this equation are inherently complicated functions of both space and time. The term "wave functions" refers to these intricate functions. Unlike real functions of time and space, Schrodinger wave functions do not possess these characteristics (Yang, 2005). These functions are intricate descriptions that depict the hidden wave-like behavior of a quantum particle.

Based on these solutions, one can generate alternative sorts of solutions. For example:

$$\Psi(x, t) = A_1\, e^{i(k_1 x - \omega_1 t)} + A_2\, e^{i(k_2 x - \omega_2 t)},$$

With

$$\hbar\omega_1 = \frac{\hbar^2 k_1^2}{2m} \quad \text{and} \quad \hbar\omega_2 = \frac{\hbar^2 k_2^2}{2m},$$

where A1 and A2 are random complex constants; this can be readily verified by directly substituting values. The most comprehensive method involves combining sinusoidal waves with all conceivable angular frequencies and wave numbers, resulting in a superposition (Umeyama, 2012).

$$\Psi(x, t) = \int_{-\infty}^{+\infty} A(k')\, e^{i(k'x-\omega' t)}\, dk' \quad \text{with} \quad \hbar\omega' = \frac{\hbar^2 k'^2}{2m}.$$

Within this superposition, A(k') is a function that can be any complex number and is dependent on k_0. This is the case if the function A(k') is characterized by a limited range of wave numbers that are centered on a positive value k (Lind et al., 2012). A wave packet of this kind is a representation of a quantum particle that moves at a speed of hk/m.

3.2.2. Wave Equation for a Particle in a Potential Energy Field

In general, the behavior of a particle that is not moving at relativistic speeds can be explained by taking into consideration the effects of a potential energy field. As a means of illustration, the movement of an electron within a hydrogen atom can be imagined as taking place within the potential energy field V(r) = −e2/4πϵ0r that is generated by the nucleus (Lévy-Leblond, 1967). According to the principles of classical physics, it is well acknowledged that when an electron is situated at a distance r from the nucleus (Bhabha, 1945), it experiences a pulling force of magnitude e2/4πϵ0r.

A wave equation was developed by Erwin Schrodinger in 1926 in order to provide a description of the behavior of quantum particles when they are in a potential energy field. When it comes to precisely characterizing the characteristics of atoms and other tiny systems (Fierz & Pauli, 1939), this equation has proven to be quite useful. An extension of the wave equation that takes into account a particle that is not subject to any external forces.

An equation that is commonly referred to as the Schrodinger equation is the one that describes the behavior of a particle (Evans, 2003) that is moving in a three-dimensional potential energy field V(r):

$$i\hbar\frac{\partial\Psi}{\partial t} = \left[-\frac{\hbar^2}{2m}\nabla^2 + V(\mathbf{r})\right]\Psi.$$

When a particle undergoes motion in a 1-D potential V(x):

$$i\hbar\frac{\partial\Psi}{\partial t} = \left[-\frac{\hbar^2}{2m}\frac{\partial^2}{\partial x^2} + V(x)\right]\Psi.$$

In situations when the potential energy remains constant, the process of determining solutions to the Schrodinger equation is rather straightforward (Emami et al., 2010). In the event that a particle is moving along the x axis and possesses a constant potential energy V_0, for example, the wave function of the particle can be defined as follows:

$$\Psi(x,t) = A\,e^{i(kx-\omega t)}$$

$$\hbar\omega = \frac{\hbar^2k^2}{2m} + V_0.$$

The particle is thus expressed by the following equation:

$$E = \frac{p^2}{2m} + V_0 \quad \text{and} \quad p = \hbar k.$$

ACTIVITY 3.1.

Objective: Understand the behavior of particles in a potential energy field using simulations.

Materials: Quantum mechanics simulation software (e.g., PhET Interactive Simulations).

Instructions:

1. Use the simulation to model a particle in a potential well.
2. Observe how the wave function and energy levels change with different potential well depths and widths.
3. Record the energy eigenvalues and corresponding wave functions for different potential well configurations.

Discussion Questions:

1. How do the energy levels change as the potential well depth increases?
2. What happens to the wave function as the width of the potential well changes?
3. How does the concept of quantization manifest in the energy levels of the particle?

SUMMARY

- In the first part of this chapter, the features of waves are investigated, and then the fundamental components of the Schrodinger equation are discussed. It is possible to define sinusoidal waves as basic solutions to wave equations. These waves are differentiated by their regular oscillations in both space and time. Additionally, the idea of linear superposition is investigated, which states that the wave function that is produced as a result of the propagation of multiple waves over the same medium is the sum of the wave functions that are produced by each individual wave. The fundamental significance of this principle is essential to the comprehension of phenomena such as interference and diffraction. The chapter also makes a distinction between waves that are dispersive and waves that are not dispersive. Dispersive waves are made up of frequency components that move at different phase velocities, which causes the wave structure to change over the course of time. Waves that are non-dispersive, on the other hand, keep their shape even as they travel over space.
- Specifically, the equations of particle waves are the focus of the second section of the discussion. A free particle's wave function must be consistent with the Schrodinger equation, which is a type of equation that is unique to free particles. This equation is used to characterize wave functions of free particles that have momentum and energy dimensions that are clearly defined. Following that, the chapter goes on to elaborate on this idea by applying it to particles that are located in potential energy fields. In doing so, the Schrodinger equation is modified to include a potential term. It is vital to have an updated equation that takes into account the influence of potential energy on the particle in order to have a complete understanding of how particles behave in a variety of potential landscapes. The purpose of this chapter is to lay forth the fundamental ideas that are required for understanding quantum systems in both unobstructed space and in the presence of potential fields.

REVIEW QUESTIONS

1. Define a sinusoidal wave and write its general mathematical form?
2. Explain the principle of linear superposition?
3. Differentiate between dispersive and non-dispersive waves?
4. What kind of solutions can be obtained from the free particle Schrödinger equation?
5. Write down the Schrödinger equation for a particle in a potential energy field?
6. Explain how the solutions to the Schrödinger equation change when a potential energy field is introduced?

MULTIPLE CHOICE QUESTIONS

1. **The principle of linear superposition states that:**
   a. The resultant wave function is the product of individual wave functions.
   b. The resultant wave function is the sum of individual wave functions.
   c. The resultant wave function is the average of individual wave functions.
   d. The resultant wave function is the difference of individual wave functions.
2. **Which type of waves maintain their shape as they propagate?**
   a. Dispersive waves
   b. Non-dispersive waves
   c. Both dispersive and non-dispersive waves
   d. Neither dispersive nor non-dispersive waves
3. **How does the Schrödinger equation for a particle in a potential energy field differ from that of a free particle?**
   a. It includes a time-independent term.
   b. It includes an additional term for potential energy.
   c. It excludes the kinetic energy term.
   d. It does not differ; they are the same.
4. **How does the Schrödinger equation relate to the concept of wave-particle duality?**
   a. It describes the particle aspect of quantum objects.
   b. It describes the wave aspect of quantum objects.
   c. It provides a framework for both wave and particle aspects of quantum objects.
   d. It only applies to particles with zero mass.
5. **Why are boundary conditions important in solving the Schrödinger equation?**
   a. They determine the kinetic energy of a particle.
   b. They define the initial position of a particle.
   c. They ensure the solutions are physically meaningful and specific to the problem.
   d. They are not important in solving the Schrödinger equation.

Answers to Multiple Choice Questions

1. (b); 2. (b); 3. (b); 4. (c); 5. (c)

REFERENCES

1. Babanin, A. V., & Rogers, W. E. (2014). Generation and limiters of rogue waves. *The International Journal of Ocean and Climate Systems, 5*(2), 39–49.
2. Belavkin, V. P., & Staszewski, P. (1992). Nondemolition observation of a free quantum particle. *Physical Review A, 45*(3), 1347.

3. Bhabha, H. J. (1945). Relativistic wave equations for the elementary particles. *Reviews of Modern Physics, 17*(2–3), 200.
4. Buchleitner, A., Delande, D., & Zakrzewski, J. (2002). Non-dispersive wave packets in periodically driven quantum systems. *Physics Reports, 368*(5), 409–547.
5. Budiyono, A., & Umeno, K. (2009). Madelung fluid model for the most likely wave function of a single free particle in a two-dimensional space with a given average energy. *Physical Review A: Atomic, Molecular, and Optical Physics, 79*(4), 042104.
6. Chraplyvy, Z. V. (1953). Reduction of relativistic two-particle wave equations to approximate forms. I. *Physical Review, 91*(2), 388.
7. Dagan, Y., & Bush, J. W. (2020). Hydrodynamic quantum field theory: The free particle. *Comptes Rendus. Mécanique, 348*(6–7), 555–571.
8. Didenkulova, I., & Pelinovsky, E. (2009). Non-dispersive traveling waves in inclined shallow water channels. *Physics Letters A, 373*(42), 3883–3887.
9. Donelan, M. A., Anctil, F., & Doering, J. C. (1992). A simple method for calculating the velocity field beneath irregular waves. *Coastal Engineering, 16*(4), 399–424.
10. Emami-Razavi, M., Bergeron, N., & Darewych, J. W. (2010). Relativistic wave equations of combined three particles and three antiparticles in scalar quantum field theory. *Journal of Physics G: Nuclear and Particle Physics, 37*(2), 025007.
11. Evans, M. W. (2003). A generally covariant wave equation for grand unified field theory. *Foundations of Physics Letters, 16*(6), 513–547.
12. Feit, M. D., Fleck Jr., J. A., & Steiger, A. (1982). Solution of the Schrödinger equation by a spectral method. *Journal of Computational Physics, 47*(3), 412–433.
13. Fierz, M., & Pauli, W. E. (1939). On relativistic wave equations for particles of arbitrary spin in an electromagnetic field. *Proceedings of the Royal Society of London. Series A. Mathematical and Physical Sciences, 173*(953), 211–232.
14. Franaszczuk, P. J., & Blinowska, K. J. (1985). Linear model of brain electrical activity—EEG as a superposition of damped oscillatory modes. *Biological Cybernetics, 53*(1), 19–25.
15. Givoli, D., & Neta, B. (2003). High-order non-reflecting boundary conditions for dispersive waves. *Wave Motion, 37*(3), 257–271.
16. Greenhow, M., & White, S. P. (1997). Optimal heave motion of some axisymmetric wave energy devices in sinusoidal waves. *Applied Ocean Research, 19*(3–4), 141–159.
17. Guerrero, J., López-Ruiz, F. F., Aldaya, V., & Cossio, F. (2011). Harmonic states for the free particle. *Journal of Physics A: Mathematical and Theoretical, 44*(44), 445307.
18. Horton, W. (1999). Drift waves and transport. *Reviews of Modern Physics, 71*(3), 735.
19. Kosevich, Y. A. (1993). Nonlinear sinusoidal waves and their superposition in anharmonic lattices. *Physical Review Letters, 71*(13), 2058.
20. Kuntz, H. L., & Blackstock, D. T. (1987). Attenuation of intense sinusoidal waves in air-saturated, bulk porous materials. *The Journal of the Acoustical Society of America, 81*(6), 1723–1731.
21. Lévy-Leblond, J. M. (1967). Nonrelativistic particles and wave equations. *Communications in Mathematical Physics, 6*, 286–311.

22. Lind, S. J., Xu, R., Stansby, P. K., & Rogers, B. D. (2012). Incompressible smoothed particle hydrodynamics for free-surface flows: A generalized diffusion-based algorithm for stability and validations for impulsive flows and propagating waves. *Journal of Computational Physics, 231*(4), 1499–1523.
23. Madsen, P. A., & Schaeffer, H. A. (2010). Analytical solutions for tsunami runup on a plane beach: Single waves, N-waves, and transient waves. *Journal of Fluid Mechanics, 645*, 27–57.
24. Misra, P. K. (2012). On the wave-particle duality of photons using the concept of a particle possessing an energy field. *Physics Essays, 25*(1), 127–132.
25. Nambu, Y. (1966). Relativistic wave equations for particles with internal structure and mass spectrum. *Progress of Theoretical Physics Supplement, 37*, 368–382.
26. Nayfeh, A. H. (1974). Sound waves in two-dimensional ducts with sinusoidal walls. *The Journal of the Acoustical Society of America, 56*(3), 768–770.
27. Nimmo, J. J., & Schief, W. K. (1997). Superposition principles associated with the Moutard transformation: An integrable discretization of a (2+1)-dimensional sine-Gordon system. *Proceedings of the Royal Society of London. Series A: Mathematical, Physical and Engineering Sciences, 453*(1957), 255–279.
28. Okanuma, S. (2006). Consideration of the fundamental characteristics of a sinusoidal-wave output inverter utilizing magnetic oscillation. *Journal of the Magnetics Society of Japan, 30*(2), 286.
29. Roberts, C. S., & Buchsbaum, S. J. (1964). Motion of a charged particle in a constant magnetic field and a transverse electromagnetic wave propagating along the field. *Physical Review, 135*(2A), A381.
30. Schleich, W. P., Greenberger, D. M., Kobe, D. H., & Scully, M. O. (2013). Schrödinger equation revisited. *Proceedings of the National Academy of Sciences, 110*(14), 5374–5379.
31. Sheremet, A., & Guza, R. T. (1999). A weakly dispersive edge wave model. *Coastal Engineering, 38*(1), 47–52.
32. Simonis, G. J., Sattler, J. P., Worchesky, T. L., & Leavitt, R. P. (1984). Characterization of near-millimeter wave materials by means of non-dispersive Fourier transform spectroscopy. *International Journal of Infrared and Millimeter Waves, 5*, 57–72.
33. Umeyama, M. (2012). Eulerian–Lagrangian analysis for particle velocities and trajectories in a pure wave motion using particle image velocimetry. *Philosophical Transactions of the Royal Society A: Mathematical, Physical and Engineering Sciences, 370*(1964), 1687–1702.
34. Winfree, A. T. (1990). Stable particle-like solutions to the nonlinear wave equations of three-dimensional excitable media. *SIAM Review, 32*(1), 1.
35. Yang, C. D. (2005). Wave-particle duality in complex space. *Annals of Physics, 319*(2), 444–470.
36. Zhang, Z., Li, B., Gao, G., & Li, M. (2021). Measurement of wave structures of non-dispersive guided waves in opaque media. *Measurement Science and Technology, 32*(11), 115601.

CHAPTER

4

Quantum Mechanics in One Dimension

LEARNING OBJECTIVES

After studying this chapter, you will be able to:

1. Understand the quantum states of a particle in a one-dimensional square-well potential.
2. Distinguish between bound states with discrete energies and unbound states that have continuous range of energies.
3. Analyze the behavior of particles incident on a square potential barrier and comprehend the phenomena of reflection and transmission.
4. Explain the concept of quantum tunneling through potential barriers.
5. Solve the Schrödinger equation for simple potential energy fields and interpret the resulting wave functions.
6. Apply the principles of quantum mechanics to model the bound and unbound states of particles in potential wells and barriers.

KEY TERMS FROM THIS CHAPTER

bound states
energy eigenvalues
quantum states
Schrödinger equation
unbound states
energy eigenfunctions
potential barrier
quantum tunneling
square-well potential
wave function

UNIT INTRODUCTION

One can gain valuable understanding of how quantum particles interact with potential energy fields by examining models that involve square barriers and square wells. In these models, solving Schrodinger equation becomes simple with basic mathematical tools, allowing us to easily determine the potential energies of particles and understand the inherent characteristics of their wave functions (Sukumar, 1985).

First, let's analyze the quantum states of a particle that is restricted within a one-dimensional square-well potential. We will demonstrate the existence of unbound states characterized by a continuous spectrum of energies, as well as the presence of bound states with discrete energies when the potential well reaches a certain depth (Jona et al., 1981).

Let's now examine a particle that encounters a square potential barrier. The nature of this encounter, which presents two potential outcomes: reflection and transmission will also be discussed. One crucial aspect to consider is the possibility of transmission, even in cases where the particle lacks the necessary energy to overcome the barrier. Put simply, we will demonstrate the phenomenon of quantum particles tunneling through potential barriers (Coutinho et al., 1997).

This chapter explores the quantum mechanics of particles in one-dimensional potential energy fields, focusing on square wells and square barriers. It covers the analysis of bound and unbound states, examining how particles behave when subjected to these potential energy configurations. By solving the Schrödinger equation for these models, the chapter demonstrates the principles of quantum tunneling and provides a clear understanding of the wave function behavior in both confined and open systems. This foundational knowledge is essential for understanding more complex quantum systems and their applications in various fields of physics.

4.1. BOUND AND UNBOUND STATES

To investigate the characteristics of bound and unbound quantum states within a simple framework, one will examine particle with mass m in a one-dimensional potential energy field.

$$V(x) = \begin{cases} \infty & \text{if } -\infty < x < 0 \\ -V_0 & \text{if } 0 < x < a \\ 0 & \text{if } a < x < \infty \end{cases} \qquad (1)$$

At x = 0 and x = a, the potential energy changes suddenly, as seen in Figure 4.1. There exists a potential well of depth , that has the potential to either confine or release the particle, alongside a solid wall at x = 0 that pushes the particle away (Baroni et al., 2013).

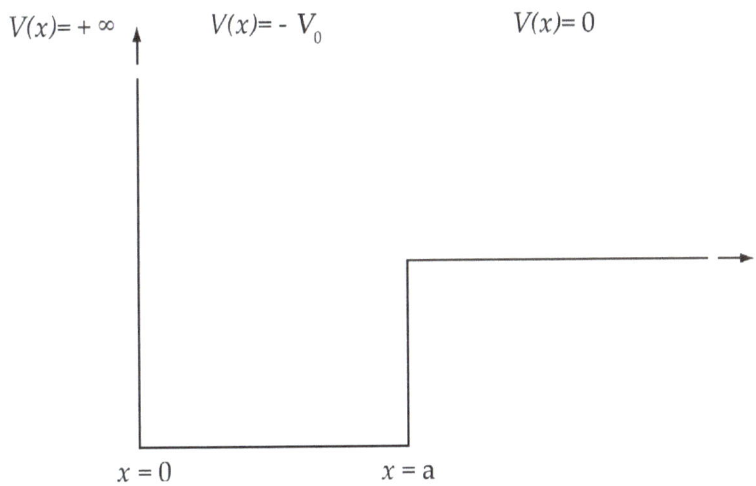

Figure 4.1. Illustration of the potential energy represented by Eqn. (1), where bound states with discrete energies and unbound states spanning continuous range of energies exist, depending on how deep the well is.

Source: Hilbert, Creative Commons License.

It should be well known how a classical particle behaves in this potential. Combining the particle's kinetic and potential energies yields its energy, K;

$$K = \frac{p^2}{2m} + V(x).$$

When the energy is negative and falls within a range from to E = 0, the particle becomes confined or trapped within the well of depth V_0. It oscillates back and forth between x = 0 and x = a, possessing a combined kinetic energy of . However, in the case of positive energy, the particle is not bound (Pearce et al., 2017). As an example, a possible scenario involves the particle approaching the potential well from positive infinity with a certain amount of kinetic energy (x = +∞.). As it progresses towards the position x = a, its kinetic energy increases to a value of . Eventually, it encounters the infinitely high potential barrier at x = 0, causing it to rebound back towards x = +∞.

The wave function describes the behavior of a quantum particle in this potential, as it satisfies the Schrodinger equation (Shegelski et al., 2008).

$$i\hbar\frac{\partial\Psi}{\partial t} = -\frac{\hbar^2}{2m}\frac{\partial^2\Psi}{\partial x^2} + V(x)\Psi. \quad (2)$$

The wave function takes the following shape when the particle has a defined energy of E.

$$\Psi(x,\ t) = \psi(x)\,e^{-iEt/\hbar}, \quad (3)$$

In the given equation, represents an eigenfunction that satisfies the energy eigenvalue equation.

$$-\frac{\hbar^2}{2m}\frac{d^2\psi}{dx^2} + V(x)\psi = E\psi. \quad (4)$$

Once this eigenvalue equation is solved and all possible energy eigenvalues and eigenfunctions are found, any quantum state of the particle in the potential can be expressed as a linear superposition of energy eigenfunctions (Sen et al., 1974).

To solve the eigenvalue equation, it is important to observe that the potential V(x) described by Eqn. (1) remains constant in three distinct areas of x: (0 < x < a), (-∞ < x < 0), and (a < x < +∞). To generate physically acceptable eigenfunctions, we will solve Eqn. (4) separately in these three locations and then combine the answers at x = 0 and x = a.

The eigenfunctions can only be as smooth as feasible due to the suddenly changes of potential V(x) at x = 0 and x = a. Specifically, we will need ψ(x) to be continuous at x = 0 and x = a to prevent the position probability density from changing suddenly in an unacceptable way. The first derivative of is therefore implied to be continuous at x = a and discontinuous at x = 0.1 by the differential Eqn. (4) with potential Eqn. (1).

Two kinds of eigenfunctions are of relevance to us. The bound state eigenfunctions and the unbound state eigenfunctions (Tanaka et al., 1999).

4.1.1. Bound States

When considering the existence of a bound state, it is worth noting that its energy lies within the range of negative values, specifically between and . Let's set , with ϵ representing the binding energy, and then look for solutions of Eqn. (4) (Avishai et al., 1991).

In the region where x is less than zero, the potential energy becomes infinitely large (. In this case, the only solution to Eqn. (4) is This indicates that the particle cannot be found in the negative x region. Within the specified region (0 < x < a), the potential energy remains constant at . Eqn. (4) takes on a specific form in this context (Marinica et al., 2008).

$$\frac{d^2\psi}{dx^2} = -k_0^2\psi, \quad \text{where} \quad E = \frac{\hbar^2 k_0^2}{2m} - V_0. \tag{5}$$

This second-order differential equation's general solution takes the following form:

$$\psi(x) = C\sin(k_0 x + \gamma),$$

where the arbitrary constants C and γ are used. In order to guarantee the continuation of at x = 0, we will set to zero (Lipkin, 1986), which will provide;

$$\psi(x) = C\sin k_0 x. \tag{6}$$

The potential energy is zero in the regionand Eqn. (4) takes on the form

$$\frac{d^2\psi}{dx^2} = \alpha^2\psi, \quad \text{where} \quad E = -\frac{\hbar^2\alpha^2}{2m}. \tag{7}$$

The general solution is;

$$\psi(x) = A\,e^{-\alpha x} + A'\,e^{+\alpha x},$$

Where the constants A and A' are arbitrary. By setting A' to zero, we can guarantee that the eigenfunction is finite at infinity and obtain a solution that decreases exponentially with x:

Remember

In a sufficiently deep square-well potential, bound states have discrete energy levels. This means that the energy of a particle in these states can only take on specific values, unlike in classical mechanics where any energy value is possible

$$\psi(x) = A\,e^{-\alpha x}. \tag{8}$$

The next step is to connect solution provided by Eqn. (6), that is applicable in range (0 < x < a), with the solution given by Eqn. (8), which is applicable in the range (a < x < +∞). As mentioned earlier, it is crucial for the eigenfunction and its first derivative to exhibit continuity at x = a. The smoothness of ψ(x) is essential (Plotnik et al., 2011).

$$C \sin k_0 a = A\,e^{-\alpha a}, \tag{9}$$

and d/dx continuity gives

$$k_0 C \cos k_0 a = -\alpha A\,e^{-\alpha a}. \tag{10}$$

Eqn. (10) divided by Eqn. (9) yields

$$k_0 \cot k_0 a = -\alpha. \tag{11}$$

The condition for a smooth connection of the functions and at x an is given by Eqn. (11). This is a non-trivial condition that can only be met when certain values for the parameters k_0 and an are taken. Once we have discovered these unique values, we can also use m to find the binding energies of the bound states (Bargmann, 1952).

In order to determine these binding energies, it is important to recognize that and are not independent variables. They are characterized by;

$$E = -\frac{\hbar^2 \alpha^2}{2m} \quad \text{and} \quad E = \frac{\hbar^2 k_0^2}{2m} - V_0.$$

Which imply that

$$\alpha^2 + k_0^2 = w^2, \quad \text{where } w \text{ is given by} \quad V_0 = \frac{\hbar^2 w^2}{2m}. \tag{12}$$

As a result, we have Eqns. (11 and 12), which are two simultaneous equations for a and . To solve these equations graphically, locate the curves' points of intersection (Kang et al., 2023).

$$\alpha = -k_0 \cot k_0 a \quad \text{and} \quad \alpha^2 + k_0^2 = w^2,$$

As illustrated in Figure 4.2.

Upon examining Figure 4.2, it becomes evident that the quantity of intersection points, and consequently bound states number, grows as depth of the well increases (Buell & Shadwick, 1995). Specifically, there are no bound states for a shallow well with

$$w < \frac{\pi}{2a}.$$

One bound state exists when;

$$\frac{\pi}{2a} < w < \frac{3\pi}{2a},$$

As well as two bound states at;

$$\frac{3\pi}{2a} < w < \frac{5\pi}{2a},$$

And so on.

We will assume potential with a well-depth parameter of , corresponding to a well of depth, in order to demonstrate the nature of the bound states (Johnson et al., 2020).

$$V_0 = \frac{2\hbar^2 \pi^2}{m a^2}.$$

Two bound states are present in this scenario: a first excited state and a ground state, each with specific binding energies.

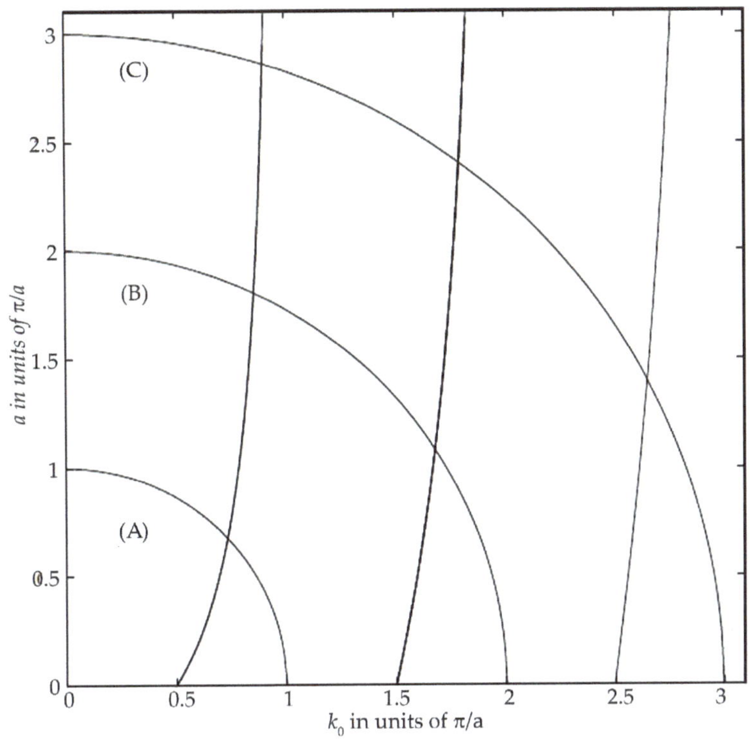

Figure 4.2. Illustration of the Graphical solution of the simultaneous equations the units of and are π/a.

Note: Three different values are assigned labels (A), (B) and (C) to represent the well-depth parameter. The values are , and , respectively. In scenario A, there is a one bound state and single point of intersection. In scenario B, there are two bound states and two points of intersection. Lastly, in scenario C, there are three bound states and three points of intersection.

Source: Leonard, Creative Commons License.

$$\epsilon_1 = 3.26\,\frac{\hbar^2\pi^2}{2ma^2} \quad \text{and} \quad \epsilon_2 = 1.17\,\frac{\hbar^2\pi^2}{2ma^2}.$$

The eigenfunctions that match to the given information are displayed in Figure 4.3. The eigenfunctions demonstrate that a restricted quantum particle can exist beyond the classical zone of confinement ($0 < x < a$). Specifically, when x is greater than 0, the eigenfunctions of bound states have non-zero values that may be determined (Gordon, 1969).

$$\psi(x) = A\,e^{-\alpha x}.$$

Therefore, probability densities for a bound particle position decrease exponentially as x enters the region where classical motion is not allowed. The parameter ⊠ and the binding energy ϵ are related through the equation . The degree of quantum penetration into the classically prohibited zone is particularly noticeable with low binding energies (Zhen et al., 2014).

4.1.2. Unbound States

By noticing the particle in Figure 4.1, one can see that it has positive energy E and is moving toward the well on the right and is reflected at x = 0. One can express the particle's energy as;

$$E = \frac{\hbar^2 k_0^2}{2m} - V_0 = \frac{\hbar^2 k^2}{2m} \qquad (13)$$

It should be noted that if the particle were subject to classical physics, its momentum would be when it was outside the well and when it was inside. However, the fundamental ideas of quantum physics largely dictate how the particle behaves (Vincent & Fortune, 1970).

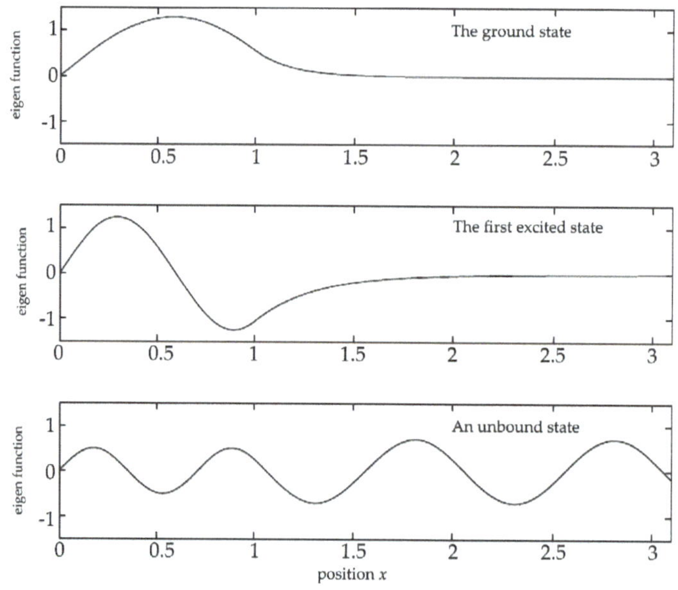

Figure 4.3. Illustration of the particle's energy eigenfunctions in a width-configured square well potential and depth . The ground state eigenfunctions, the first state, and an unbound state with energy are shown. The eigenfunction of the unbound state has been arbitrarily normalized.

Source: Fredimend, Creative Commons License.

The wave function is expressed in the form of;

$$\Psi(x, t) = \psi(x)\, e^{-iEt/\hbar}$$

where the eigenvalue Eqn. (4) is satisfied for all values of x by the eigenfunction . The method we used to discover the bound-state eigenfunctions will be applied to find these eigenfunctions as well: we will locate solutions to Eqn. (4) in each of the three regions of x, connect these solutions at x = O and x = a. First, since the potential energy is infinite in the area, the eigenfunction is zero (Rolfs & Azuma, 1974).

Furthermore, inside the area where the potential energy is , the eigenvalue equation for the unbound state has a similar structure to the eigenvalue equation for a bound state, as described in Eqn. (5). The solution may be determined using Eqn. (6), which is as follows:

$$\psi(x) = C \sin k_0 x. \qquad (14)$$

Moreover, in the area where the potential energy is zero , the eigenvalue equation can be represented as follows (Smith et al., 2015).

$$\frac{d^2\psi}{dx^2} = -k^2\psi, \qquad (15)$$

and here is the solution

$$\psi(x) = D \sin(kx + \delta). \qquad (16)$$

The phase shift is referred to as δ.

Now, we can proceed by merging the solution obtained from Eqn. (14) with the solution obtained from Eqn. (16). It is crucial to ensure the continuity of the eigenfunction and its first derivative at x = a, as mentioned earlier. The continuity of yields certain outcomes (Tobi & Bahar, 2005).

$$C \sin k_0 a = D \sin(ka + \delta) \qquad (17)$$

and dψ/dx continuity results in

$$k_0 C \cos k_0 a = kD \cos(ka + \delta). \qquad (18)$$

Eqn. (18) divided by Eqn. (17) yields

$$k_0 \cot k_0 a = k \cot(ka + \delta). \qquad (19)$$

The equation establishes the requirement for an effortless relationship between the functions and at x = a. Upon examining the corresponding equation for the eigenfunctions of the bound states, denoted as Eqn. (11), we discovered that connection is only smooth when energy assumes specific discrete values. Unbound states do not follow this pattern. In this context, we have the freedom to select any value for the energy E. We can then determine the values of k and using Eqn. (13), and subsequently utilize Eqn. (19) to calculate phase shift d. Therefore, given each positive energy E (Schult et al., 1989), it is always possible to discover a continuous eigenfunction in the following manner:

$$\psi(x) = \begin{cases} 0 & \text{if } -\infty < x < \\ C\sin(k_0 x) & \text{if } 0 < x < a \\ D\sin(kx + \delta) & \text{if } a < x < \infty. \end{cases} \quad (20)$$

Undulations with wave numbers and k are present in regions where a classical particle would have momenta and . Figure 4.3 shows the oscillations, or undulations, of the eigenfunction of a free particle. The energy of the particle is given by , and it is limited to a potential well with a depth of .

The eigenfunction Eqn. (20) can be reformulated to show the phenomenon of wave reflection by the potential well. To identify the incident and reflected components, we represent the eigenfunction using complex exponentials. Given the use of;

$$\sin\theta = \frac{e^{+i\theta} - e^{-i\theta}}{2i}$$

And if two additional constants are added

$$A_0 = -\frac{C}{2i} \quad \text{and} \quad A = -\frac{D e^{-2i\delta}}{2i},$$

We obtain

$$\psi(x) = \begin{cases} A_0 e^{-ik_0 x} - A_0 e^{+ik_0 x} & \text{if } 0 < x < a \\ A e^{-ikx} - A e^{2i\delta} e^{+ikx} & \text{if } a < x < \infty. \end{cases} \quad (21)$$

Inside the well, we can observe a wave oscillating with a wave number of .

Waves with a wave number of k are visible entering into and exiting out of the well, but there is a 2δ phase difference between them. The waves that are shown describe the reflection of a particle with momentum k by the barrier at x = 0 and by the potential well between x = 0 and x = a. Since the incoming and reflected waves have the same intensity, as indicated by ., the reflection is perfect. The significance of the phase shift will become apparent shortly (Bohlen et al., 1997).

Even while terms like "incoming," "outgoing," and "traveling back and forth" have been used, the eigenfunction provided by Eqn. (21) represents a quantum state that, because it has a defined energy K, does not show any observable time dependency. We require a non-stationary quantum state of undetermined energy in order to adequately characterize the dynamics of particle reflection (Sen et al., 1974). Such a state can be represented by an incoming wave packet of the following form in the region

$$\Psi_i(x, t) = \int_0^{\infty} c(E') e^{-i(E't+p'x)/\hbar}\, dE', \quad \text{where} \quad E' = \frac{p'^2}{2m},$$

And a wave packet that is emitted

$$\Psi_f(x, t) = \int_0^{\infty} c(E') e^{-i(E't-p'x)/\hbar}\, e^{2i\delta(E')}\, dE'.$$

The function represents the energy probability amplitude, where is the chance that the particle's energy lies between E' and E' + dE.' If the function is highly concentrated, it means that E' is equal to E, where E represents the energy of the outgoing particle. The momentum of the particle is given by , and its velocity is . In this particular case (Darden et al., 1973), it can be demonstrated that

$$\Psi_i(x,\ t) \propto F(t + x/v) \tag{22}$$

And

$$\Psi_f(x,\ t) \propto F(t - x/v - 2\hbar \mathrm{d}\delta/\mathrm{d}E), \tag{23}$$

The form of the wave packets entering and exiting is determined by function F. The formulas show the relationship between the phase shift's energy dependence, δ(E), and the reflection process's time delay (Holl et al., 2022). One way to approximate this delay is by;

$$\tau \approx 2\hbar \frac{\mathrm{d}\delta}{\mathrm{d}E}. \tag{24}$$

The time delay for a particle reflected by the potential shown in Figure 4.1 is negative because the particle accelerates as it enters the well at .

4.1.3. General Implications

This section examines a particle within the potential energy field depicted in Figure 4.1. We have chosen this fundamental potential due to its compatibility with basic solution approaches for the Schrodinger equation and its ability to generate readily observable waveforms. The specifics of the findings obviously only apply to this specific potential (Heisenberg, 2019), but the broad aspects of atomic, nuclear, and particle physics are as follows:

a. In classically permitted regions, wave functions undulate, whereas in classically banned regions, they fall off exponentially.

b. When potentials are sufficiently attractive, they give rise to bound states with distinct energies.

c. A particle's wave function undergoes a phase shift upon being scattered or reflected by a potential, which may be connected to a time delay that takes place during the scattering process. Initially, a particle can have an infinite range of energies (Guaní et al., 2010).

4.2. BARRIER PENETRATION

One of the key characteristics of a quantum particle is its capacity to traverse and tunnel through an area that is traditionally considered inaccessible. In the previous section, it was determined that the wave function of a bound particle extends beyond the limited area of a classical bound particle. This section will demonstrate the phenomenon of particle tunneling through potential barriers. In order to maintain mathematical simplicity (Di et al., 2008), we will examine a particle within a basic potential energy field characterized by following form:

$$V(x) = \begin{cases} 0 & \text{if } -\infty < x < 0 \\ V_B & \text{if } 0 < x < a \\ 0 & \text{if } a < x < +\infty. \end{cases} \tag{25}$$

In Figure 4.4, there is a square barrier with a height VB that separates two regions: (-∞ < x < 0) and

A classical particle approaching this barrier from the left would be transmitted if its energy was above and reflected if it was below. We will show that the result of a quantum particle encountering the barrier is not known; it could be transmitted or reflected. In particular, we will calculate the probability that the particle will be transferred even if its energy is less than and demonstrate that this is possible (Goodwin & Clark, 2005).

The wave function , which satisfies the Schrodinger equation, characterizes the dynamics of a particle with mass m in the potential V(x).

$$i\hbar\frac{\partial\Psi}{\partial t} = -\frac{\hbar^2}{2m}\frac{\partial^2\Psi}{\partial x^2} + V(x)\Psi.$$

In order to explain the behavior of an uncertain encounter with the barrier, we are looking for a wave function that represents a particle approaching the barrier and the potential outcomes of reflection and transmission. The wave function should provide a probability pulse that represents a particle approaching the barrier prior to the encounter (Martins et al., 2012).

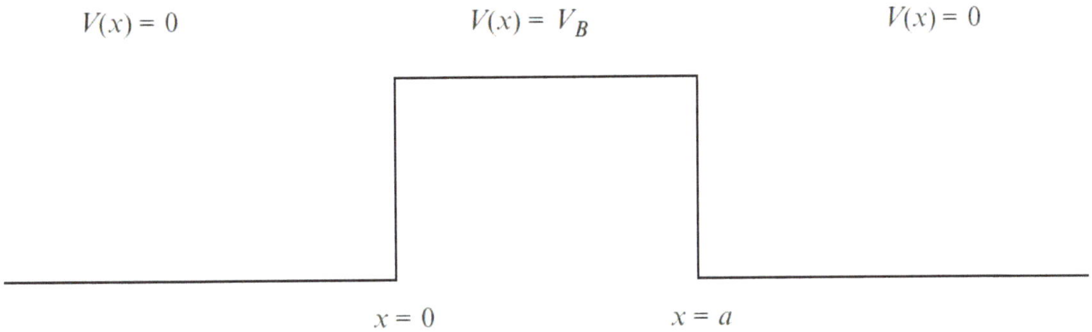

Figure 4.4. The potential barrier given by Eqn. (25) which will be used to illustrate quantum mechanical tunneling.

Source: Arfk:en, G. B, Creative Commons License.

Following the interaction, there will be two distinct pulses of probability. One pulse corresponds to probability of a particle being reflected, while the other pulse represents the probability of a particle being transmitted. According to the conventional understanding of the wave function, the possibilities of reflection and transmission remain until the particle is observed. During this occurrence, the wave function undergoes collapse, resulting in the realization of one of the two choices. The probability of these options are determined by the magnitudes of the reflected and transmitted pulses (Brink & Takigawa, 1977).

This dynamic description implies that we must address an issue in the field of time-dependent quantum physics. A non-stationary state is used to describe the quantum state of an arriving particle and the potential outcomes of transmission and reflection. A state of unknown energy is characterized by a linear combination of energy eigenfunctions. Here, there exists a range of potential energies and the wave function takes on a specific shape (Patzelt et al., 2012).

$$\Psi(x,\ t) = \int c(E')\psi_{E'}(x)\,\mathrm{e}^{-iE't/\hbar}\,\mathrm{d}E',$$

where an eigenfunction with energy E' and the proper boundary conditions is represented by (x). is the chance that the particle has energy between E' and E' + dE,' and the function c(E') represents an energy probability amplitude. When an energy particle E encounters a barrier and the function is sharply peaked at E' = E, the wave function is composed of a localized wave packet (Singh et al., 2020).

Time-dependent quantum mechanics is not needed to calculate the probabilities of reflection and transmission, but it is essential to comprehend the uncertain interaction with a potential barrier. Now, we will show how time-independent quantum mechanics can be used to calculate these probabilities.

4.2.1. Stationary State Analysis of Transmission and Reflection

When the fluctuations in the potential energy V(x) are much larger than the uncertainty in the particle's energy, one can analyze a stationary state with a known energy to determine the probabilities of reflection and transmission (Leavens & Aers, 1989). The wave function represents the state in question.

$$\psi(x,t) = \psi_E(x)e^{-iEt/\hbar}, \tag{26}$$

The function is an eigenfunction with energy K that satisfies the eigenvalue equation.

$$-\frac{\hbar^2}{2m}\frac{d^2\psi_E}{dx^2} + V(x)\psi_E = E\psi_E. \tag{27}$$

The potential V(x) shown in Figure 4.4 is a basic form, making it simple to identify an eigenfunction that characterizes incoming, reflected, and transmitted waves. The process involves finding the solutions to Eqn. (27) in various x-regions, then smoothly merging the results (Omar & Kamel, 2003).

In the case where the barrier is located to the left, the potential energy V(x) is set to zero and the eigen-function obeys a specific differential equation.

$$\frac{d^2\psi_E}{dx^2} = -k^2\psi_E, \quad \text{where} \quad E = \frac{\hbar^2 k^2}{2m}. \tag{28}$$

The equation that represents the intensity of an incident wave () and the reflected wave is

$$\psi_E(x) = A_I e^{+ikx} + A_R e^{-ikx}. \tag{29}$$

The eigenfunction's form within the barrier is determined by the energy level of the particle in relation to the barrier. When the energy is greater than the potential barrier , the region between 0 and an is considered a classically allowed region . In this region, the behavior of the eigenfunction is determined by certain governing principles (Ngo et al., 2015).

$$\frac{d^2\psi_E}{dx^2} = -k_B^2\psi_E, \quad \text{where} \quad E = \frac{\hbar^2 k_B^2}{2m} + V_B. \tag{30}$$

The generic solution oscillates with wave number and involves two arbitrary constants.

$$\psi_E(x) = Ae^{+ik_Bx} + A'e^{-ik_Bx}. \tag{31}$$

The region $(0 < x < a)$ is conventionally banned when E < . The eigenfunction in this case is controlled by;

$$\frac{d^2\psi_E}{dx^2} = \beta^2\psi_E, \quad \text{where} \quad E = -\frac{\hbar^2\beta^2}{2m} + V_B, \tag{32}$$

And the general solution is;

$$\psi_E(x) = Be^{-\beta x} + B'e^{+\beta x}, \tag{33}$$

Where B and B' are arbitrary constants.

The potential energy is once again 0 on the right side of the barrier (Koutserimpas & Fleury, 2019). The eigenfunction fulfills Eqn. (28) and represents the solution for a transmitted wave with an intensity of .

$$\psi_E(x) = A_T e^{+ikx}. \tag{34}$$

The intensities of the reflected and transmitted waves can be found by smoothly joining these solutions at x = 0 and x = a. Specifically, we can obtain formulas for the ratios;

$$R = \frac{|A_R|^2}{|A_I|^2} \quad \text{and} \quad T = \frac{|A_T|^2}{|A_I|^2}. \tag{35}$$

These ratios represent probabilities since the probability of finding a particle at x is proportional to R represents the probability that the particle is reflected, and T represents the probability that it is transmitted (Chen & Liao, 2015). The sum of these probabilities equals one, i.e.,

$$R + T = 1. \tag{36}$$

Focusing on a particle whose energy is below the barrier, we will obtain a rough equation for the transmission probability that proves to be helpful in situations where the barrier is large. Stated differently, we will determine the probability of tunneling through a large obstruction (Leavens et al., 1995).

4.2.2. Tunneling Through Wide Barriers

The eigenfunction for a particle with energy below the barrier is given by

$$\psi_E(x) = \begin{cases} A_I\, e^{+ikx} + A_R\, e^{-ikx} & \text{if } -\infty < x < 0 \\ B\, e^{-\beta x} + B'\, e^{+\beta x} & \text{if } 0 < x < a \\ A_T\, e^{+ikx} & \text{if } a < x < \infty, \end{cases} \tag{37}$$

Using the values of constants , , , B and B' ensuring that and are continuous at x = 0 and x = a (Cerezo et al., 2006). For continuity to exist at x = 0;

$$A_I + A_R = B + B' \quad \text{and} \quad ikA_I - ikA_R = -\beta B + \beta B'.$$

as well as continuity at x = a call for

$$B\,e^{-\beta a} + B'\,e^{+\beta a} = A_T\,e^{ika} \quad \text{and} \quad -\beta B\,e^{-\beta a} + \beta B'\,e^{+\beta a} = ikA_T\,e^{ika}.$$

Our goal is to derive the tunneling probability expression. In order to achieve this, we will define the constants and in terms of B. Initially, continuity at x = 0 is used to provide;

$$2ikA_I = -(\beta - ik)B + (\beta + ik)B'. \tag{38}$$

Secondly, continuity is used at x = a to provide

$$A_T\,e^{ika} = \frac{2\beta}{(\beta - ik)}B\,e^{-\beta a} \quad \text{and} \quad B' = B\,e^{-2\beta a}\frac{(\beta + ik)}{(\beta - ik)}. \tag{39}$$

Obtaining the tunneling probability and expressing both and in terms of B is now simple. T = |AT |2/|AI |2.

In order to avoid the tedious calculations using algebra, we will simplify the situation by supposing that the particle comes over a broad barrier where the exponential function (-2βa) is much smaller than the length 1 (. When this condition is met (Rampello et al., 2019), the value of constant B' is significantly smaller than B, allowing us to make an approximation of Eqn. (38) as follows:

$$2ikA_I \approx -(\beta - ik)B.$$

By integrating this equation with Eqn. (39), we derive;

$$A_T\,e^{ika} \approx -\frac{4ik\beta\,e^{-\beta a}}{(\beta - ik)^2}A_I.$$

This suggests that the probability of tunneling can be approximated by;

$$T \approx \left[\frac{16k^2\beta^2}{(\beta^2 + k^2)^2}\right] e^{-2\beta a}.$$

If we rewrite this equation using

$$k = \frac{\sqrt{2mE}}{\hbar} \quad \text{and} \quad \beta = \frac{\sqrt{2m(V_B - E)}}{\hbar},$$

We obtain

$$T \approx \left[\frac{16E(V_B - E)}{V_B^2}\right] e^{-2\beta a}. \quad (40)$$

When << l, this approximation for the tunneling probability is true. In this scenario, the exponentially decaying term dominates the wave function inside the barrier, and the chance of tunneling through a barrier of width an is proportional to . One of the key aspects of the tunneling formula Eqn. (40) is its exponential dependence on the penetration parameter β and the width of the barrier, a. This dependency has significant practical implications (Gajek et al., 2007).

We will use two processes to demonstrate the significance of barrier penetration: one involves tunneling electrons, and the other involves tunneling protons (Jauho & Jonson, 1989).

4.2.3. Tunneling Electrons

According to the photoelectric effect, the least amount of energy required to remove one electron from a metal surface is typically a few electron volts. This energy is required because electrons in a metal are located in a desirable potential energy field that becomes stronger in the direction of the metal's surface, producing a potential step that is a few electron volts higher than the energy level of the metal's most energetic electrons (Wang et al., 2005).

Two areas with low potential energy that are divided by a potential barrier when two metal surfaces are brought close to one another resemble the one shown in Figure 4.4.

However, this barrier does not impede the movement of electrons over the space between the surfaces. Electrons are quantum particles capable of tunneling over a barrier with a probability determined by Eqn. (40), or roughly estimated as

$$T \approx e^{-2\beta a} \quad \text{with} \quad \beta = \frac{\sqrt{2m_e(V_B - E)}}{\hbar}. \quad (41)$$

The expression is determined by various factors, including the gap width between the metal surfaces , the electron mass, the barrier height at the metal surface, and the energy of the most energetic electrons in the metal (Uiberacker et al., 2007).

Eqn. (41) indicates that the tunneling probability is highly dependent on the gap, exhibiting a strong sensitivity. Indeed, if the gap an undergoes a tiny shift Δa, the fractional change in the tunneling probability can be expressed as

$$\frac{\Delta T}{T} = -2\beta \, \Delta a. \quad (42)$$

In order to demonstrate the numerical importance of this equation, we will examine a standard scenario where an electron must be ejected from the metal with 4 eV. This suggests that the penetration parameter is provided by and that the height of the barrier is approximately 4 eV higher than the electron energy E (Qian et al., 2018).

$$\beta = \frac{\sqrt{2m_e(V_B - E)}}{\hbar} \approx 10^{10}\,\text{m}^{-1}. \tag{43}$$

Did you know?

Quantum tunneling is a phenomenon where particles can pass through a potential barrier that they classically should not be able to cross. This effect is fundamental in many physical processes, including nuclear fusion in stars and the operation of tunnel diodes in electronics.

By replacing this value for β in Eqn. (42), we can demonstrate the remarkable sensitivity of the electron tunneling probability. For example, a little variation of 0.001 nm in the space between the surfaces leads to a quantifiable 2% alteration in the probability (Foster et al., 1988).

The scanning tunneling microscope utilizes the high sensitivity of electron tunneling. This instrument utilizes a sharp metal probe that is placed in close proximity to a surface being examined. The spacing is reduced to a size that encourages the movement of electrons between the probe and the surface by a process called tunneling. In addition, a potential difference is generated between the probe and the surface, leading to a directional movement of electrons. When the probe is manipulated or scanned across the surface, tiny atomic properties will result in noticeable fluctuations in the electron tunneling current. The scanning tunneling microscope can generate a highly accurate map that shows the positions of individual atoms on the surface (Gull et al., 1993).

4.2.4. Tunneling Protons

The center of the sun consists of a densely charged gas consisting of electrons, protons, and light atomic nuclei, all coexisting at an extremely high temperature of around K. Protons and other light nuclei often collide, occasionally merging and releasing energy that is eventually emitted as sunlight from the solar surface. We will be focusing on two protons approaching one another close to the sun's center in order to comprehend the problems associated with the production of solar thermonuclear energy (Salna et al., 2016). Their thermal kinetic energy in the ionized gas are of the order of;

$$E \approx kT \approx 1\,\text{keV}.$$

The two protons' separation determines their mutual potential energy. The repulsive Coulomb potential dominates the potential energy at large separation u, as seen in Figure 4.5.

$$V(r) = \frac{e^2}{4\pi\epsilon_0 r}.$$

However, at short distances when the value of u is similar to the range of nuclear forces, denoted by , the potential energy becomes attractive (Harmony, 1972). The overall outcome is a Coulomb barrier that increases to an energy level of around 1 million electron volts (MeV) when the distance between the objects is around 2 ot (2 fm).

Therefore, as protons come close to each other in the core of the sun, they do so with energies on the scale of kiloelectron volts (keV) and they face a Coulomb barrier that is measured in megaelectron volts (MeV). According to classical physics (Brougham et al., 1999), the concept of the distance of closest approach, denoted as , is well-defined and may be determined using the following equation:

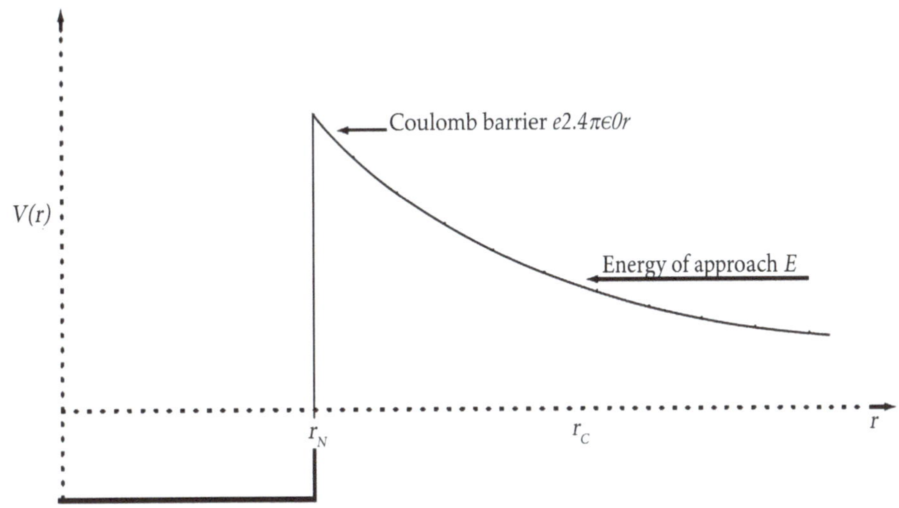

Figure 4.5. A schematic representation of two protons with a separation of u and their potential energy V (u). Protons may fuse to create deuteron with the separation is less than due to high nuclear attraction. The coulomb barrier is around 1 MeV high and prevents nuclear fusion. For protons with an energy of approach equal to K, the classical distance of closest approach is .

Source: Fetter, A. L, Creative Commons License.

$$E = \frac{e^2}{4\pi\epsilon_0 r_C}. \qquad (44)$$

Given the considerable disparity in size between this distance and the range of nuclear forces , it seems impossible that a close encounter and subsequent nuclear fusion would occur. Upon initial observation, it is obvious that the sun lacks the necessary heat to emit light (Maji et al., 2007).

Thermonuclear fusion in the sun and other stars occurs due to the quantum nature of protons, which allows them to pass past Coulomb barriers. The time-independent

Schrödinger equation describes the behavior of two protons with an energy of approach E. The eigenfunction satisfies this equation (Benoit et al., 1998).

$$\left[-\frac{\hbar^2}{2\mu}\nabla^2 + V(r)\right]\psi(\mathbf{r}) = E\psi(\mathbf{r}), \tag{45}$$

where μ equals mp/2, the decreased mass of two protons, and V(r) is the potential shown in Figure 4.4 The appropriate eigenfunction for two low-energy protons is angularly independent and has the

$$\psi(\mathbf{r}) = \frac{u(r)}{r}, \tag{46}$$

$$-\frac{\hbar^2}{2\mu}\frac{d^2u}{dr^2} + V(r)u = E. \tag{47}$$

We will refrain from attempting to answer this problem. Alternatively, we will utilize the findings from our analysis of a one-dimensional barrier to formulate a reasonable expression for the eigenfunction. Subsequently, we will make an approximation to determine the probability of tunneling over a Coulomb barrier (Robertson & Lawrence, 1981).

Let's start by examining a three-dimensional barrier that has a constant height VB and a width equal to the difference between . In this scenario, the function u(r) exhibits exponential decay within the classically prohibited zone as r decreases, and its expression is provided as follows:

$$u(r) \propto e^{\beta r},$$

Where β is given by;

$$E = -\frac{\hbar^2\beta^2}{2\mu} + V_B.$$

The probability of proton tunneling from point to can be approximated by the ratio of the to This probability is expressed as:

$$T \approx |\ \exp[-\beta(r_C - r_N)]\ |^2. \tag{48}$$

Let us now examine the Coulomb barrier shown in Figure 4.5. The eigenfunction within the classically restricted zone (where) is once again approximately determined by

$$u(r) \propto e^{\beta r},$$

nonetheless, as β is now dependent on u, given by;

$$E = -\frac{\hbar^2\beta^2}{2\mu} + \frac{e^2}{4\pi\epsilon_0 r}.$$

The probability of the two protons tunneling from tois now essentially described by a generalized version of the Eqn. 4.48 (Trixler, 2013).

$$T \approx \left| \exp\left[- \int_{r_N}^{r_C} \beta \, dr \right] \right|^2. \quad (49)$$

Assuming that the radius of , we can simplify the integral in Eqn. (49) by inserting . This substitution leads to the following result:

$$T \approx \exp\left[-\left(\frac{E_G}{E} \right)^{1/2} \right] \quad (50)$$

The relative energy of the protons is denoted as E, and KG is defined as follows:

$$E_G = \left(\frac{e^2}{4\pi\epsilon_0\hbar c} \right)^2 2\pi^2\mu c^2. \quad (51)$$

The energy is referred to as the Gamow energy, with a specific value of 493 keV.

Now, we can calculate the probability of two protons successfully tunneling through the Coulomb barrier, which typically prevents them from coming close to each other, during a collision in the core of the sun (Henini et al., 1989). When we substitute a typical thermal energy of K = 1 keV into Eqn. (50), we can obtain the desired result.

$$T \approx \exp[-22] \approx 3 \times 10^{-10}.$$

Therefore, it is extremely rare for protons to successfully penetrate the Coulomb barrier when they collide near the center of the sun, with a probability of approximately one in 3 billion. Furthermore, when these particles come together, there is an exceptional chance for fusion to occur, resulting in the release of immense thermonuclear energy (Hernández et al., 2003).

In practice, stars undergo gradual changes in their temperature, ensuring that the average thermal energy of nuclei remains significantly lower than the Coulomb barrier. The process of fusion occurs at a rate that is directly related to the probability of particles tunneling. Due to the extremely low probability, the process of fusion occurs at a slow rate, resulting in the nuclear fuel lasting for an incredibly long period of time.

ACTIVITY 4.1

Objective: To understand the concept of quantum tunneling through a hands-on simulation.

Materials Needed:

Computer with internet access, online quantum mechanics simulation tool (e.g., PhET Interactive Simulations).

Instructions:

a. **Introduction:**

- Provide a brief overview of quantum tunneling and its importance in quantum mechanics.

b. Simulation Setup:

Open the PhET Interactive Simulations website and navigate to the "Quantum Tunneling and Wave Packets" simulation.

c. Exploration:

- Set up a potential barrier in the simulation.
- Launch a wave packet towards the barrier and observe the behavior of the wave function.
- Vary the height and width of the barrier and note the changes in transmission and reflection probabilities.

d. Discussion:

- Discuss why the wave packet can sometimes penetrate the barrier even when its energy is less than the barrier height.
- Relate the observations to real-world phenomena like electron tunneling in semiconductors.

SUMMARY

- This chapter delves into the quantum mechanics of particles in one-dimensional potential energy fields, focusing on square wells and square barriers. By solving the Schrödinger equation for these models, the chapter explores the quantum states of particles, distinguishing between bound states with discrete energies and unbound states with continuous energy ranges.
- It then examines particles encountering a square potential barrier, showing that quantum mechanics allows for the possibility of tunneling through barriers, a phenomenon impossible in classical mechanics. Through mathematical derivations and graphical solutions, the chapter provides a comprehensive understanding of how particles behave under these potential energy configurations, emphasizing the practical implications in fields like atomic, nuclear, and particle physics.

REVIEW QUESTIONS

1. Explain the differences between bound states and unbound states in a one-dimensional square-well potential.
2. Describe the concept of quantum tunneling and its significance in quantum mechanics.
3. How is the Schrödinger equation solved for a particle in a one-dimensional square-well potential?
4. What are the boundary conditions applied to the wave function at the edges of the potential well and barrier?
5. Discuss the implications of quantum tunneling in real-world applications.
6. Analyze the behavior of the wave function in the classically forbidden region of a potential barrier.
7. Explain how the transmission and reflection probabilities are calculated for a particle encountering a square potential barrier.

MULTIPLE CHOICE QUESTIONS

1. What type of states exist in a sufficiently deep square-well potential?

a. Both bound states with discrete energies and unbound states with continuous energies

b. Only bound states with discrete energies

c. Only unbound states with continuous energies

d. Neither bound states nor unbound states

2. What phenomenon allows a quantum particle to penetrate a potential barrier?

a. Classical reflection

b. Quantum tunneling

c. Elastic scattering

d. Inelastic scattering

3. What happens to the wave function of a particle in the classically forbidden region of a potential barrier?

a. It oscillates

b. It remains constant

c. It falls off exponentially

d. It increases exponentially

4. Which term describes the possible outcome when a particle encounters a potential barrier?

a. Absorption

b. Reflection

c. Emission

d. Annihilation

5. What characterizes bound states in a quantum mechanical system?

a. Continuous range of energies

b. Discrete energy levels

c. Infinite range of energies

d. Variable energy levels

6. What type of potential is considered in this chapter for analyzing quantum mechanical behavior?

a. Harmonic potential

b. Square-well potential

c. Linear potential

d. Exponential potential

Answers to Multiple Questions

1. (a); 2. (b); 3. (c); 4. (c); 5. (b); 6. (b)

REFERENCES

1. Avishai, Y., Bessis, D., Giraud, B. G., & Mantica, G. (1991). Quantum bound states in open geometries. Physical Review B, 44(15), 8028.
2. Bargmann, V. (1952). On the number of bound states in a central field of force. Proceedings of the National Academy of Sciences, 38(11), 961–966.
3. Baroni, S., Navrátil, P., & Quaglioni, S. (2013). Unified ab initio approach to bound and unbound states: No-core shell model with continuum and its application to ^{7}He. Physical Review C—Nuclear Physics, 87(3), 034326.

4. Benoit, M., Marx, D., & Parrinello, M. (1998). Tunnelling and zero-point motion in high-pressure ice. Nature, 392(6673), 258–261.
5. Bohlen, H. G., Von Oertzen, W., Stolla, T., Kalpakchieva, R., Gebauer, B., Wilpert, M., & Massey, T. N. (1997). Study of weakly bound and unbound states of exotic nuclei with binary reactions. Nuclear Physics A, 616(1–2), 254–261.
6. Brink, D. M., & Takigawa, N. (1977). Barrier penetration effects in the semi-classical theory of elastic scattering between complex nuclei. Nuclear Physics A, 279(1), 159–188.
7. Brougham, D. F., Caciuffo, R., & Horsewill, A. J. (1999). Coordinated proton tunnelling in a cyclic network of four hydrogen bonds in the solid state. Nature, 397(6716), 241–243.
8. Buell, W. F., & Shadwick, B. A. (1995). Potentials and bound states. American Journal of Physics, 63(3), 256–258.
9. Cerezo, A., Petford-Long, A. K., Larson, D. J., Pinitsoontorn, S., & Singleton, E. W. (2006). The formation mechanism of aluminum oxide tunnel barriers. Journal of Materials Science, 41, 7843–7852.
10. Chen, T., & Liao, Z. (2015). Steady-state responses and wave propagations in the periodic plate. Archive of Applied Mechanics, 85, 837–854.
11. Coutinho, F. A. B., Nogami, Y., & Perez, J. F. (1997). Generalized point interactions in one-dimensional quantum mechanics. Journal of Physics A: Mathematical and General, 30(11), 3937.
12. Darden, S. E., Sen, S., Hiddleston, H. R., Aymar, J. A., & Yoh, W. A. (1973). Study of (d, p) reactions to unbound states of ^{13}C and ^{17}O. Nuclear Physics A, 208(1), 77–92.
13. Di, L., Kerns, E. H., & Carter, G. T. (2008). Strategies to assess blood–brain barrier penetration. Expert Opinion on Drug Discovery, 3(6), 677–687.
14. Foster, J. S., Frommer, J. E., & Arnett, P. C. (1988). Molecular manipulation using a tunnelling microscope. Nature, 331(6154), 324–326.
15. Gajek, M., Bibes, M., Fusil, S., Bouzehouane, K., Fontcuberta, J., Barthélémy, A., & Fert, A. (2007). Tunnel junctions with multiferroic barriers. Nature Materials, 6(4), 296–302.
16. Goodwin, J. T., & Clark, D. E. (2005). In silico predictions of blood-brain barrier penetration: Considerations to "keep in mind". Journal of Pharmacology and Experimental Therapeutics, 315(2), 477–483.
17. Gordon, R. G. (1969). New method for constructing wavefunctions for bound states and scattering. The Journal of Chemical Physics, 51(1), 14–25.
18. Guaní-Guerra, E., Santos-Mendoza, T., Lugo-Reyes, S. O., & Terán, L. M. (2010). Antimicrobial peptides: General overview and clinical implications in human health and disease. Clinical Immunology, 135(1), 1–11.
19. Gull, S., Lasenby, A., & Doran, C. (1993). Electron paths, tunnelling, and diffraction in the spacetime algebra. Foundations of Physics, 23(10), 1329–1356.
20. Harmony, M. D. (1972). Quantum mechanical tunnelling in chemistry. Chemical Society Reviews, 1(2), 211–228.

21. Heisenberg, L. (2019). A systematic approach to generalizations of General Relativity and their cosmological implications. Physics Reports, 796, 1–113.
22. Henini, M., Leadbeater, M. L., Alves, E. S., Eaves, L., & Hughes, O. H. (1989). Ballistic transport in resonant tunnelling devices with wide quantum wells. Journal of Physics: Condensed Matter, 1(18), 3025.
23. Hernández, E., Jáuregui, A., & Mondragón, A. (2003). Jordan blocks and Gamow-Jordan eigenfunctions associated with a degeneracy of unbound states. Physical Review A, 67(2), 022721.
24. Holl, M., Lindberg, S., Heinz, A., Kondo, Y., Nakamura, T., Tostevin, J. A., & Samurai21 Collaboration. (2022). Border of the island of inversion: Unbound states in Ne^{29}. Physical Review C, 105(3), 034301.
25. Jauho, A. P., & Jonson, M. (1989). Tunnelling through a time-modulated barrier—Relation to tunnelling times. Journal of Physics: Condensed Matter, 1(45), 9027.
26. Johnson, C. W., Launey, K. D., Auerbach, N., Bacca, S., Barrett, B. R., Brune, C. R., & Zhang, X. (2020). White paper: From bound states to the continuum. Journal of Physics G: Nuclear and Particle Physics, 47(12), 123001.
27. Jona-Lasinio, G., Martinelli, F., & Scoppola, E. (1981). New approach to the semiclassical limit of quantum mechanics: I. Multiple tunnelings in one dimension. Communications in Mathematical Physics, 80, 223–254.
28. Kang, M., Liu, T., Chan, C. T., & Xiao, M. (2023). Applications of bound states in the continuum in photonics. Nature Reviews Physics, 5(11), 659–678.
29. Koutserimpas, T. T., & Fleury, R. (2019). Coupled-mode theory for stationary and nonstationary resonant sound propagation. Wave Motion, 89, 221–231.
30. Leavens, C. R., & Aers, G. C. (1989). Dwell time and phase times for transmission and reflection. Physical Review B, 39(2), 1202.
31. Leavens, C. R., Iannaccone, G., & McKinnon, W. R. (1995). On the approach to the stationary-state-scattering limit within Bohmian mechanics. Physics Letters A, 208(1–2), 17–24.
32. Lipkin, H. J. (1986). A model-independent approach to multiquark bound states. Physics Letters B, 172(2), 242–247.
33. Maji, K., Mondal, C. K., & Bhattacharyya, S. P. (2007). Tunnelling time and tunnelling dynamics. International Reviews in Physical Chemistry, 26(4), 647–670.
34. Marinica, D. C., Borisov, A. G., & Shabanov, S. V. (2008). Bound states in the continuum in photonics. Physical Review Letters, 100(18), 183902.
35. Martins, I. F., Teixeira, A. L., Pinheiro, L., & Falcao, A. O. (2012). A Bayesian approach to in silico blood-brain barrier penetration modeling. Journal of Chemical Information and Modeling, 52(6), 1686–1697.
36. Ngo, T., Lwin, M., & Santoso, S. (2015). Steady-state analysis and performance of low frequency AC transmission lines. IEEE Transactions on Power Systems, 31(5), 3873–3880.

37. Omar, A. S., & Kamel, A. H. (2003). Steady-state analysis of signal transmission in evanescent channels. IEEE Transactions on Antennas and Propagation, 51(5), 957–964.
38. Patzelt, A., Lademann, J., Richter, H., Darvin, M. E., Schanzer, S., Thiede, G., & Hauser, M. (2012). In vivo investigations on the penetration of various oils and their influence on the skin barrier. Skin Research and Technology, 18(3), 364–369.
39. Pearce, N. M., Krojer, T., & Von Delft, F. (2017). Proper modelling of ligand binding requires an ensemble of bound and unbound states. Acta Crystallographica Section D: Structural Biology, 73(3), 256–266.
40. Plotnik, Y., Peleg, O., Dreisow, F., Heinrich, M., Nolte, S., Szameit, A., & Segev, M. (2011). Experimental observation of optical bound states in the continuum. Physical Review Letters, 107(18), 183901.
41. Qian, H., Hsu, S. W., Gurunatha, K., Riley, C. T., Zhao, J., Lu, D., & Liu, Z. (2018). Efficient light generation from enhanced inelastic electron tunnelling. Nature Photonics, 12(8), 485–488.
42. Rampello, S., Fantera, L., & Masini, L. (2019). Efficiency of embedded barriers to mitigate tunnelling effects. Tunnelling and Underground Space Technology, 89, 109–124.
43. Robertson, G. N., & Lawrence, M. C. (1981). Analysis of proton tunnelling in KDP using a realistic proton potential. Journal of Physics C: Solid State Physics, 14(31), 4559.
44. Rolfs, C., & Azuma, R. E. (1974). Interference effects in ^{12}C (p, γ) ^{13}N and direct capture to unbound states. Nuclear Physics A, 227(2), 291–308.
45. Salna, B., Benabbas, A., Sage, J. T., van Thor, J., & Champion, P. M. (2016). Wide-dynamic-range kinetic investigations of deep proton tunnelling in proteins. Nature Chemistry, 8(9), 874–880.
46. Schult, R. L., Ravenhall, D. G., & Wyld, H. W. (1989). Quantum bound states in a classically unbound system of crossed wires. Physical Review B, 39(8), 5476.
47. Sen, S., Darden, S. E., Hiddleston, H. R., & Yoh, W. A. (1974). Study of the (d, p) reaction to bound and unbound states of ^{19}O. Nuclear Physics A, 219(3), 429–449.
48. Shegelski, M. R., Hnybida, J., Friesen, H., Lind, C., & Kavka, J. (2008). Tunneling of a diatomic molecule with unbound states in one dimension. Physical Review A—Atomic, Molecular, and Optical Physics, 77(3), 032702.
49. Singh, M., Divakaran, R., Konda, L. S. K., & Kristam, R. (2020). A classification model for blood brain barrier penetration. Journal of Molecular Graphics and Modelling, 96, 107516.
50. Smith, J. K., Baumann, T., Brown, J., DeYoung, P. A., Frank, N., Hinnefeld, J., & Williams, S. J. (2015). Selective population of unbound states in ^{10}Li. Nuclear Physics A, 940, 235–241.
51. Sukumar, C. V. (1985). Supersymmetric quantum mechanics of one-dimensional systems. Journal of Physics A: Mathematical and General, 18(15), 2917.

52. Tanaka, N., Suzuki, Y., Varga, K., & Lovas, R. G. (1999). Unbound states by analytic continuation in the coupling constant. Physical Review C, 59(3), 1391.
53. Tobi, D., & Bahar, I. (2005). Structural changes involved in protein binding correlate with intrinsic motions of proteins in the unbound state. Proceedings of the National Academy of Sciences, 102(52), 18908–18913.
54. Trixler, F. (2013). Quantum tunnelling to the origin and evolution of life. Current Organic Chemistry, 17(16), 1758–1770.
55. Uiberacker, M., Uphues, T., Schultze, M., Verhoef, A. J., Yakovlev, V., Kling, M. F., & Krausz, F. (2007). Attosecond real-time observation of electron tunnelling in atoms. Nature, 446(7136), 627–632.
56. Vincent, C. M., & Fortune, H. T. (1970). New method for distorted-wave analysis of stripping to unbound states. Physical Review C, 2(3), 782.
57. Wang, W., Lee, T., & Reed, M. A. (2005). Electron tunnelling in self-assembled monolayers. Reports on Progress in Physics, 68(3), 523.
58. Zhen, B., Hsu, C. W., Lu, L., Stone, A. D., & Soljačić, M. (2014). Topological nature of optical bound states in the continuum. Physical Review Letters, 113(25), 257401.

CHAPTER

5

Angular Momentum

LEARNING OBJECTIVES

After studying this chapter, you will be able to:

1. Understand orbital angular momentum and its Cartesian components in quantum mechanics.
2. Know angular momentum operators, their commutation relations, and eigenvalue properties.
3. Understand matrix representation of angular momentum, eigenstates, and diagonal matrices in quantum mechanics.
4. Understand geometric representation of angular momentum vector and its components.
5. Understand experimental evidence and theoretical basis of spin angular momentum.
6. Describe matrix representation and properties of spin 1/2 using Pauli matrices.

KEY TERMS FROM THIS CHAPTER

angular momentum
eigenstates
orbital angular momentum
quantum numbers
spin states
commutation relations
matrix representation
Pauli matrices
spin angular momentum
Stern-Gerlach experiment

UNIT INTRODUCTION

In physics, angular momentum is linear momentum's rotational equivalent. Other names for it include rotating momentum and moment of momentum. The total angular momentum of a closed system remains constant due to conservation, making it a significant physical quantity. The conservation of angular momentum is an essential principle in physics, as it incorporates both the direction and magnitude of this significant physical property. The useful properties of motorcycles and bicycles, rifled bullets, flying discs, and gyroscopes are all a result of the principle of conservation of angular momentum. The formation of spirals in hurricanes and the high rotational rates of neutron stars can be linked to the principle of conservation of angular momentum. Typically, the concept of conservation places constraints on the potential motion of a system, yet it does not provide a singular solution (Leader & Lorcé, 2014). The position vector of the particle (with respect to a chosen origin) multiplied by the momentum vector ($p = mv$ in Newtonian physics) yields a pseudovector, which is the mathematical representation of the three-dimensional angular momentum of a point particle in classical mechanics. Unlike linear momentum, which is independent of the choice of origin, angular momentum is affected by the choice of origin (Yao & Padgett, 2011).

A key principle of physics is angular momentum, which asserts that the total angular momentum of a system is equivalent to the sum of the angular momentum of its constituent parts. In physics, one of the most important characteristics of angular momentum is its extensiveness. The angular momentum density across the whole body must be considered when calculating the total angular momentum of a continuous rigid body or fluid.

The angular momentum per unit volume as the volume gets closer to zero is represented by this density (Herr & Popovic, 2008). When there is no external torque operating on the system, angular momentum is preserved, just like linear momentum. Like force, torque is the unit of measurement for the rate at which angular momentum changes. A system's total torque applied to it always equals its net external torque. Put more simply, there is always zero as the total of all internal torques in a system. In rotational dynamics, this principle is comparable to Newton's third law of motion (Padgett et al., 2004).

Therefore, in the event of a closed system with no net external torque, the system's total torque must be zero, and as a result, the system's total angular momentum stays constant. Angular momentum is equally significant in both classical mechanics and quantum mechanics. It is especially beneficial for examining the behavior of systems that undergo motion due to central potentials or spherically symmetric,, as the orbital angular momenta of these systems remain constant (Dixon, 1970).

One of the fundamental principles of Bohr's model of the hydrogen atom is the quantization of angular momentum. In this model, the electron moves within the Coulomb potential of the proton, which is a central potential. Moreover, angular momentum is crucial in explaining molecular rotations, the movement of electrons in atoms, and the mobility of nucleons in nuclei. Therefore, a thorough understanding of the quantum theory of angular momentum is necessary to analyze and investigate molecular, atomic, and nuclear systems (Franke et al., 2004). The general theory of angular momentum will be covered in this chapter. The angular momentum operator's many properties will be examined, and then we'll focus on determining its eigenvalues and eigenstates. The eigenvectors and eigenvalues related to the orbital and spin angular momenta will ultimately be found using this mathematical approach (Miao et al., 2016).

5.1. ORBITAL ANGULAR MOMENTUM

When it comes to classical physics, one important concept to consider is the angular momentum of a particle. This quantity is defined based on the particle's momentum (p) and position (r).

$$\vec{L} = \vec{r} \times \vec{p} = (yp_z - zp_y)\hat{i} + (zp_x - xp_z)\hat{j} + (xp_y - yp_x)\vec{k}. \quad (1)$$

By replacing the momentum and position operators (Chen et al., 2019), and $\vec{\ }$, one can obtain the orbital angular momentum operator

$$\hat{\vec{L}} = \hat{\vec{R}} \times \hat{\vec{P}} = -i\hbar\hat{\vec{R}} \times \vec{\nabla}. \quad (2)$$

The components of vector in Cartesian coordinates are:

$$\hat{L}_x = \hat{Y}\hat{P}_z - \hat{Z}\hat{P}_y = -i\hbar\left(\hat{Y}\frac{\partial}{\partial z} - \hat{Z}\frac{\partial}{\partial y}\right),$$
$$\hat{L}_y = \hat{Z}\hat{P}_x - \hat{X}\hat{P}_z = -i\hbar\left(\hat{Z}\frac{\partial}{\partial x} - \hat{X}\frac{\partial}{\partial z}\right),$$
$$\hat{L}_z = \hat{X}\hat{P}_y - \hat{Y}\hat{P}_x = -i\hbar\left(\hat{X}\frac{\partial}{\partial y} - \hat{Y}\frac{\partial}{\partial x}\right). \quad (3)$$

$$\hat{L}_x = \hat{Y}\hat{P}_z - \hat{Z}\hat{P}_y = -i\hbar\left(\hat{Y}\frac{\partial}{\partial z} - \hat{Z}\frac{\partial}{\partial y}\right),$$
$$\hat{L}_y = \hat{Z}\hat{P}_x - \hat{X}\hat{P}_z = -i\hbar\left(\hat{Z}\frac{\partial}{\partial x} - \hat{X}\frac{\partial}{\partial z}\right),$$
$$\hat{L}_z = \hat{X}\hat{P}_y - \hat{Y}\hat{P}_x = -i\hbar\left(\hat{X}\frac{\partial}{\partial y} - \hat{Y}\frac{\partial}{\partial x}\right). \quad (4)$$

$$\hat{L}_x = \hat{Y}\hat{P}_z - \hat{Z}\hat{P}_y = -i\hbar\left(\hat{Y}\frac{\partial}{\partial z} - \hat{Z}\frac{\partial}{\partial y}\right),$$
$$\hat{L}_y = \hat{Z}\hat{P}_x - \hat{X}\hat{P}_z = -i\hbar\left(\hat{Z}\frac{\partial}{\partial x} - \hat{X}\frac{\partial}{\partial z}\right),$$
$$\hat{L}_z = \hat{X}\hat{P}_y - \hat{Y}\hat{P}_x = -i\hbar\left(\hat{X}\frac{\partial}{\partial y} - \hat{Y}\frac{\partial}{\partial x}\right). \quad (5)$$

It is obvious that angular momentum is not possible in a one-dimensional space. It is important to note that, its square, and its components, are all Hermitian.

$$\hat{\vec{L}}^2 = \hat{L}_x^2 + \hat{L}_y^2 + \hat{L}_z^2, \quad (6)$$

5.2. GENERAL FORMALISM OF ANGULAR MOMENTUM

Angular momentum operator defined by its three components , , will be explored in depth. These components satisfy the following commutation relations (Penrose, 1982).

$$[\hat{J}_x, \hat{J}_y] = i\hbar \hat{J}_z, \qquad [\hat{J}_y, \hat{J}_z] = i\hbar \hat{J}_x, \qquad [\hat{J}_z, \hat{J}_x] = i\hbar \hat{J}_y, \tag{7}$$

or equivalently by;

$$\hat{\vec{J}} \times \hat{\vec{J}} = i\hbar \hat{\vec{J}}. \tag{8}$$

, cannot be simultaneously diagonalized, meaning they do not share eigenstates, because they do not mutually commute. The angular momentum squared;

$$\hat{\vec{J}}^2 = \hat{J}_x^2 + \hat{J}_y^2 + \hat{J}_z^2, \tag{9}$$

As it is a scalar operator, it commutes with , :

$$[\hat{\vec{J}}^2, \hat{J}_k] = 0, \tag{10}$$

where k represents y, z, and x. For example, in the scenario where k = x, we have

$$\begin{aligned} [\hat{\vec{J}}^2, \hat{J}_x] &= [\hat{J}_x^2, \hat{J}_x] + \hat{J}_y[\hat{J}_y, \hat{J}_x] + [\hat{J}_y, \hat{J}_x]\hat{J}_y + \hat{J}_z[\hat{J}_z, \hat{J}_x] + [\hat{J}_z, \hat{J}_x]\hat{J}_z \\ &= \hat{J}_y(-i\hbar \hat{J}_z) + (-i\hbar \hat{J}_z)\hat{J}_y + \hat{J}_z(i\hbar \hat{J}_y) + (i\hbar \hat{J}_y)\hat{J}_z \\ &= 0, \end{aligned} \tag{11}$$

In other words, [,] = , and [,] = since [] = 0. It should be noted that the eigenvalues of the operators , ,, and are real since they are Hermitian (Bliokh et al., 2017).

5.3. ANGULAR MOMENTUM MATRIX REPRESENTATION

The formalism discussed in the previous section is universal and not reliant on any specific representation. There exist numerous methods for representing their eigenstates and angular momentum operators. The matrix representation of angular momentum will be discussed in this part. Column vectors and square matrices, respectively, will be used to represent eigenkets and operators. This is accomplished by enlarging states and operators inside a separate foundation. Later, we shall explore the method of representing orbital angular momentum in the position representation (Atkins & Dobson, 1971).

The set of their common eigenstates, , can be used as a basis since and commute; this basis is discrete, orthonormal, and complete. The ortho-normalization requirement for this base is provided by Eqn. (12) for a given value of (Moses 1970), and the completeness condition is defined by;

$$\sum_{m=-j}^{+j} | j, m\rangle\langle j, m | = \hat{I}, \tag{12}$$

where represents unit matrix. The operators and are diagonal in the basis formed by their joint eigenstates.

$$\begin{aligned} \langle j', m' | \hat{\vec{J}}^2 | j, m\rangle &= \hbar^2 j(j+1)\delta_{j',j}\delta_{m',m}, \\ \langle j', m' | \hat{J}_z | j, m\rangle &= \hbar m \delta_{j',j}\delta_{m',m}. \end{aligned} \tag{13}$$

Therefore, in the eigen basis, the matrices representing and are diagonal, with their diagonal elements equal to and respectively.

Given the non-commutativity of the operators and , their representation in the basis is characterized by non-diagonal matrices (Moore & Fleming, 2006).

$$\boxed{\langle j', m' | \hat{J}_\pm | j, m\rangle = \hbar\sqrt{j(j+1) - m(m\pm 1)}\, \delta_{j',j}\delta_{m',m\pm 1}.} \tag{14}$$

We can infer the matrices of and from Eqns. (15) and (16):

$$\langle j',\, m' \mid \hat{J}_x \mid j,\, m\rangle = \frac{\hbar}{2}\left[\sqrt{j(j+1)-m(m+1)}\,\delta_{m',\,m+1} + \sqrt{j(j+1)-m(m-1)}\,\delta_{m',\,m-1}\right]\delta_{j',\,j}, \tag{15}$$

$$\langle j',\, m' | \hat{J}_y \mid j,\, m\rangle = \frac{\hbar}{2i}\left[\sqrt{j(j+1)-m(m+1)}\,\delta_{m',\,m+1} - \sqrt{j(j+1)-m(m-1)}\,\delta_{m',\,m-1}\right]\delta_{j',\,j}. \tag{16}$$

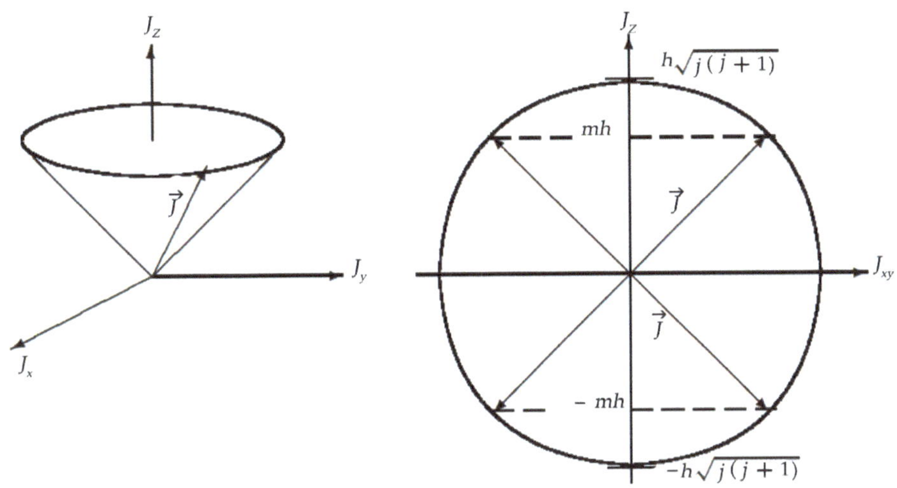

Figure 5.1. Angular momentum geometric representation.

Source: A. C. Phillips, Creative Commons License.

Did you know?

Angular momentum in quantum mechanics can be either orbital (associated with the motion of particles around a fixed point) or spin (intrinsic to particles themselves, akin to their "spin").

5.4. GEOMETRICAL REPRESENTATION OF ANGULAR MOMENTUM

This involves relationship between the angular momentum and its z-component, which has the following geometric representation (Hsiang & Straume, 2008). The total angular momentum for a fixed value of j can be represented by a vector whose z-component is and whose length, as shown in Figure 5.1, is given by ˆ. and are distinct, hence the sole well-defined part of their total, , is contained in the xy plane (Rakitzis & Zare, 1999).

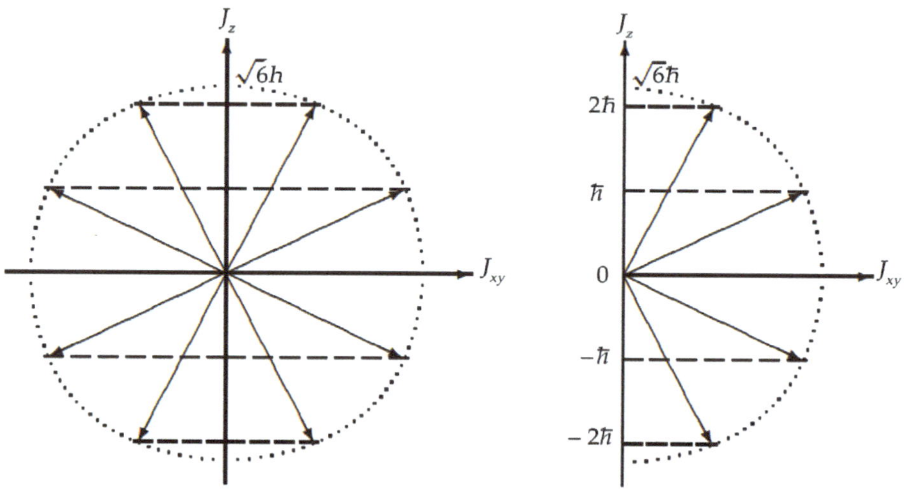

Figure 5.2. Graphical representation of the angular momentum j = 2 for the state | 2, m with. The radius of the circle is .

Source: D. J. Sandiford, Creative Commons License.

can be conceptualized, in classical terms, as a vector whose endpoint is located on a circle of radius spinning around the surface of a half-angle cone so that its projection down the z-axis is always $m\hbar$.

$$\theta = \cos^{-1}\left(\frac{m}{\sqrt{j(j+1)}}\right),$$

(17)

It is important to observe that the quantum number m has a limited range of values, specifically from -. This limitation leads to the quantization of the angle θ, meaning that θ can only take on a discrete set of 1 values (Aquilanti et al., 2009).

$$\theta = \cos^{-1}\left(\frac{-j}{\sqrt{j(j+1)}}\right), \cos^{-1}\left(\frac{-j+1}{\sqrt{j(j+1)}}\right), \ldots, \cos^{-1}\left(\frac{j-1}{\sqrt{j(j+1)}}\right)$$
$$\cos^{-1}\left(\frac{j}{\sqrt{j(j+1)}}\right). \qquad (18)$$

Remember

Cones and vectors can be used to graphically depict angular momentum. Only discrete values, or the quantum numbers , can be projected onto any axis by the angular momentum vector.

The projection of on both the x and y axes averages out to zero because all orientations of J;ˆ on the cone's surface are equally likely (Hehl & McCrea, 1986).

$$\langle \hat{J}_x \rangle = \langle \hat{J}_y \rangle = 0, \qquad (19)$$

N is represented by the symbol .

For example, the graphical representation for the scenario is displayed in Figure 5.2. As stated in Eqn. (4.18), there is only a discrete set of values that θ can take. For in this instance when , there are only five possible values for the angle θ, and they are provided (Keetels et al., 2008) by;

$$\theta = -35.26^{\circ} - 65.91^{\circ}, 90^{\circ}, 65.91^{\circ}, 35.26^{\circ}. \qquad (20)$$

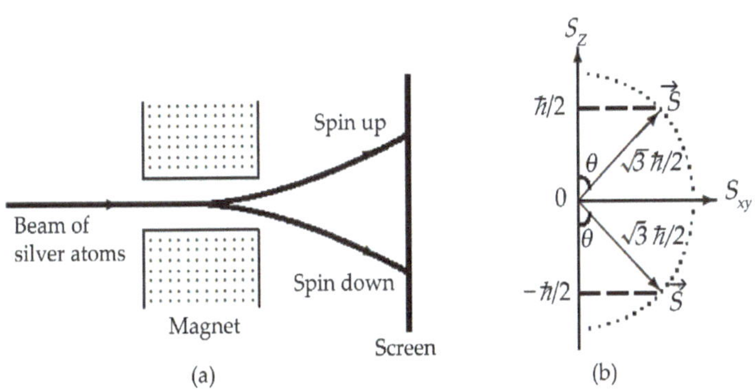

Figure 5.3. Illustration of (a) The Stern-Gerlach experiment shows that a beam of silver atom separates into two separate parts that represent spin-up and spin-down when it travels through an inhomogeneous magnetic field. (b) A representation of spin shows that the point of ; is located on a circle with radius ||= meaning that there are only two possible values for its projection on the z-axis takes only two values, ± /2, with = 54.7°.

Source: F. Mandl, Creative Commons License.

5.5. SPIN ANGULAR MOMENTUM

5.5.1. Spin Experimental Evidence

Silver (Ag) atoms were used by (Barke et al., 2006) to empirically verify the presence of spin. There are 47 electrons in silver; 46 of them make up a charge distribution that is spherically symmetric, and the 47th electron resides in a 5s orbital. Due to the placement of the fifth shell electron in a 5s state, the silver atom's overall orbital angular momentum in the ground state would be zero, denoted as .

The Stern-Gerlach experiment involves the passage of a silver atom beam through a magnetic field that is not uniform (inhomogeneous). Assuming that the field is aligned in the z-direction, we should expect observing a symmetrical continuous band on the screen, centered on the undisturbed direction, $z = 0$, based on classical expectations (Pan et al., 2001).

On the other hand, Schrödinger's wave theory would predict that beam would split into an odd (discrete) number of components if atoms had an orbital angular momentum of l. If the electron in the fifth shell were in a 5p state (), three spots would show up on the screen. There would only be one point, though, if the atoms in the beam were in their ground state ().

However, in terms of behavior, the beam does not follow either Schrödinger's wave theory or classical physics expectations. Rather, it divides into two independent parts, as Figure 5.3a shows. This outcome was also noted for ground state hydrogen atoms , where splitting is not expected (Vaara et al., 2002).

In 1925, Goudsmit and Uhlenbeck proposed a solution to this puzzle. They suggested that the electron has an intrinsic angular momentum, separate from its orbital angular momentum, which is not related to its spatial degrees of freedom. In drawing a parallel to the Earth's movement, where it orbits the Sun and also spins on its axis, it is possible to view the electron, or any other tiny particle, as having a similar internal spinning motion (Emmerich et al., 1985).

Spin angular momentum is a provocative moniker for this intrinsic degree of freedom. It is important to note that the electron is a point-like or structureless particle. Therefore, caution should be exercised while trying to establish a connection between the electron's spin and an internal spinning motion. A particle's spin angular momentum is independent of its spatial degrees of freedom. There is no classical equivalent for the spin, an intrinsic degree of freedom that is exclusively related to quantum mechanics. In contrast to orbital angular momentum, differential operators are unable to adequately characterize spin (Chen et al., 2020).

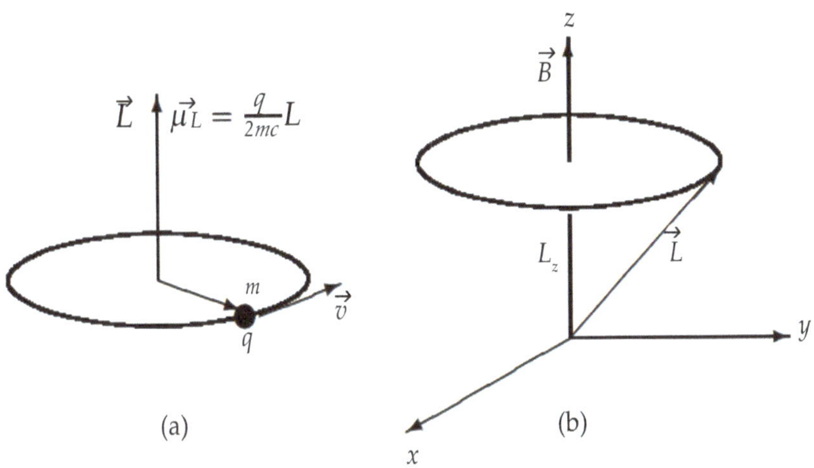

Figure 5.4. Illustration of (a) The orbital magnetic dipole moment with a positive charge q. (b) The application of an external magnetic field causes the orbital magnetic moment to undergo precession.

Source: A. C. Phillips, Creative Commons License.

Did you know?

Spin angular momentum is quantized in units of ħ/2 (reduced Planck's constant divided by 2), leading to the classification of particles as either integer spin (bosons) or half-integer spin (fermions).

The classical theory of electromagnetism states that the orbital movement of a charged particle produces an orbital magnetic dipole moment:

$$\vec{\mu}_L = \frac{q}{2mc}\vec{L}, \tag{21}$$

In Figure 5.4a, it is observed that when the charge q is positive, the vectors and $\vec{}$will be parallel to each other. On the other hand, in the case of a negatively charged particle such as an electron , the magnetic dipole moment , and the direction of the orbital angular momentum will be opposite (Mydosh, 2015).

By adopting a classical analysis and envisioning the electron as a rotating spherical charge, we can deduce an inherent or spin magnetic dipole moment The conventional derivation of ES is highly inadequate as the electron cannot be regarded as a rotating sphere. In reality, it has been discovered that the electron's spin magnetic moment is twice as large as its classical representation (Mayer, 1950). While the traditional derivation of the spin magnetic moment, similar to the orbital magnetic moment, is not possible, it can still be postulated via analogy with Eqn. (22).

$$\vec{\mu}_S = ü_s \frac{e}{ü_e} \vec{}\qquad (22)$$

where is known as the gyromagnetic ratio of the electron or the Landé factor; Dirac's relativistic theory of the electron may be used to determine this factor, which has an experimental value of .

An electron's intrinsic dipole moment is affected by the application of a magnetic field . The force's direction and strength are affected by the dipole's relative orientation to the field and the magnetic field's uniformity (Benner, 1978). A processional motion of around is produced by this force, which tends to align along (Figure 5.4b).

An electron will travel in the direction of an increasing field, for example, if is parallel to ; on the other hand, an electron will move in the direction of a decreasing field, if is antiparallel to When hydrogen-like atoms, like silver, are in their ground state, their orbital angular momentum is zero. As a result, the electron's spin alone accounts for the whole dipole moment of the atom (Leyder et al., 2007).

The deflection of the atomic beam is determined by the orientation of the electron's spin. Based on experimental observations, it has been observed that the beam splits into two segments, suggesting that the spin of the electron can only exist in two specific orientations in relation to the magnetic field: either antiparallel or parallel.

Similar to an analogy, two quantum numbers determine a particle's orbital angular momentum. The spin angular momentum properties are defined by the orbital number l and the azimuthal number . Two quantum numbers, s and its projection m_s down the z-axis (the direction of the magnetic field), where further describe the spin angular momentum (Atxitia et al., 2010).

Given that the Stern-Gerlach experiment detected only two components, it follows that 2s + 1 = 2. The electron's quantum numbers are determined by and

Every fundamental particle in existence have a distinct spin. Certain particles possess spins that are either integers (0, 1, 2, etc.) or half-odd-integers . For example, pi mesons have a spin of 0, photons have a spin of 1, and so on. On the other hand, electrons, protons, and neutrons have a spin of 1, whereas deltas have a spin of 1/2, and so forth. Fermions, like as quarks, electrons, protons, and neutrons, are particles with half-odd-integer spins. On the other hand, bosons, such pions, photons, and gravitons, have integer spins (Maeno et al., 1999).

In addition to verifying and quantifying spin, the Stern-Gerlach experiment provides several other significant applications of quantum physics. First, it supports the discrete nature of the microphysical universe postulated by the quantum theory by demonstrating that a beam divides into a distinct set of components rather than a continuous band (Light, 2013).

The Stern-Gerlach experiment is a highly valuable approach for preparing a quantum state. To generate a beam consisting of atoms with spin-up orientation, one can achieve this by directing an unpolarized beam through a magnet that has varying magnetic field strength. Subsequently, we can choose gather the necessary component while disregarding or obstructing the remaining atoms (Zamanian et al., 2010).

The Stern-Gerlach experiment is also applicable in determining the overall angular momentum of an atom. When l is not equal to zero, the total angular momentum (J) is determined by combining the orbital angular momentum (L) and the spin angular momentum (S): $\vec{}$.

5.5.2. General Theory of Spin

In theory, angular momentum is the same as in the theory of spin. The spin, like the vector angular momentum , is defined by a vector operator . The components of this operator, , , have the same commutation relations as , .

$$[\hat{S}_x, \hat{S}_y] = i\hbar\hat{S}_z, \qquad [\hat{S}_y, \hat{S}_z] = i\hbar\hat{S}_x, \qquad [\hat{S}_z, \hat{S}_x] = i\hbar\hat{S}_y. \tag{)23}$$

Furthermore, the vectors and exhibit commutativity, indicating that they share eigenvectors (Gerratt, 1971).

$$\hat{\vec{S}}^2 \mid s, m_s\rangle = \hbar^2 s(s+1) \mid s, m_s\rangle, \qquad \hat{S}_z \mid s, m_s\rangle = \hbar m_s \mid s, m_s\rangle, \tag{24}$$

The set of values for In a similar manner, we possess

$$\hat{S}_{\pm} \mid s, m_s\rangle = \hbar\sqrt{s(s+1) - m_s(m_s \pm 1)} \mid s, m_s \pm 1\rangle, \tag{25}$$

Where and

The spin states constitute a foundation that is both orthonormal and complete (Andreev, 2015).

$$\langle \hat{S}_x^2\rangle = \langle \hat{S}_y^2\rangle = \frac{1}{2}(\langle \hat{\vec{S}}^2\rangle - \langle \hat{S}_z^2\rangle) = \frac{\hbar^2}{2}\left[s(s+1) - m_s^2\right], \tag{26}$$

Where denotes

$$\langle s', m_s' \mid s, m_s\rangle = \delta_{s',s}\delta_{m_s', m_s}, \qquad \sum_{m_s=-s}^{s} \mid s, m_s\rangle\langle s, m_s \mid = I, \tag{27}$$

Where I represents the identity matrix.

5.5.3. The Pauli Matrices and Spin 1/2

The quantum number for a particle with spin 1/2 can only have two values:nd . The particle can exist in two states (Dyall, 1994): = = , and ,

$$\hat{\vec{S}}^2 \left|\frac{1}{2}, \pm\frac{1}{2}\right\rangle = \frac{3}{4}\hbar^2 \left|\frac{1}{2}, \pm\frac{1}{2}\right\rangle, \qquad \hat{S}_z \left|\frac{1}{2}, \pm\frac{1}{2}\right\rangle = \pm\frac{\hbar}{2} \left|\frac{1}{2}, \pm\frac{1}{2}\right\rangle. \tag{28}$$

Therefore, the spin can be visually shown, as seen in Figure 5.3b, by a vector with a magnitude of ||= . The tip of this vector rests on a circle with a radius of , which rotates over the surface of a cone with a half-angle (Florkowski et al., 2018).

$$\theta = \cos^{-1}\left(\frac{|m_s|}{\sqrt{s(s+1)}}\right) = \cos^{-1}\left(\frac{\hbar/2}{\sqrt{3}\hbar/2}\right) = \cos^{-1}\left(\frac{1}{\sqrt{3}}\right) = 54.73^{\circ}. \tag{29}$$

There are only two possible values for the z-axis projection of : , which represent spin-up and spin-down, respectively (Leibfried & Wineland, 2018).

Now, let's examine the spin matrix representation. The operators and within the basis can be represented by the following matrices using Eqns. (27) and (28) (Asenjo et al., 2012).

$$\vec{S}^2 = \begin{pmatrix} \left\langle \frac{1}{2},\frac{1}{2}\middle|\vec{S}^2\middle|\frac{1}{2},\frac{1}{2}\right\rangle & \left\langle \frac{1}{2},\frac{1}{2}\middle|\vec{S}^2\middle|\frac{1}{2},-\frac{1}{2}\right\rangle \\ \left\langle \frac{1}{2},-\frac{1}{2}\middle|\vec{S}^2\middle|\frac{1}{2},\frac{1}{2}\right\rangle & \left\langle \frac{1}{2},-\frac{1}{2}\middle|\vec{S}^2\middle|\frac{1}{2},-\frac{1}{2}\right\rangle \end{pmatrix} = \frac{3\hbar^2}{4}\begin{pmatrix} 1 & 0 \\ 0 & 1 \end{pmatrix} \tag{30}$$

$$\hat{S}_z = \frac{\hbar}{2}\begin{pmatrix} 1 & 0 \\ 0 & -1 \end{pmatrix} \tag{31}$$

It is possible to deduce the matrices of and from Eqn. (30 and 31):

$$\hat{S}_+ = \hbar\begin{pmatrix} 0 & 1 \\ 0 & 0 \end{pmatrix}, \qquad \hat{S}_- = \hbar\begin{pmatrix} 0 & 0 \\ 1 & 0 \end{pmatrix}, \tag{32}$$

and since

$$\hat{S}_x = \frac{\hbar}{2}\begin{pmatrix} 0 & 1 \\ 1 & 0 \end{pmatrix}, \qquad \hat{S}_y = \frac{\hbar}{2}\begin{pmatrix} 0 & -i \\ i & 0 \end{pmatrix}. \tag{33}$$

The two-element column matrices known as spinors are used to express the joint eigenvectors of and :

$$\left|\frac{1}{2},\frac{1}{2}\right\rangle = \begin{pmatrix} 1 \\ 0 \end{pmatrix}, \qquad \left|\frac{1}{2},-\frac{1}{2}\right\rangle = \begin{pmatrix} 0 \\ 1 \end{pmatrix}. \tag{34}$$

It is easy to confirm that these eigenvectors form a comprehensive basis (Wang et al., 2014);

$$\sum_{m_s=-\frac{1}{2}}^{\frac{1}{2}} \left|\frac{1}{2}, m_s\right\rangle\left\langle\frac{1}{2}, m_s\right| = \begin{pmatrix} 0 \\ 1 \end{pmatrix}(0\ \ 1) + \begin{pmatrix} 1 \\ 0 \end{pmatrix}(1\ \ 0) = \begin{pmatrix} 1 & 0 \\ 0 & 1 \end{pmatrix}, \tag{35}$$

and orthonormal;

$$\left\langle \frac{1}{2}, \frac{1}{2} \middle| \frac{1}{2}, \frac{1}{2} \right\rangle = (1 \ 0)\begin{pmatrix} 1 \\ 0 \end{pmatrix} = 1,$$
$$\left\langle \frac{1}{2}, -\frac{1}{2} \middle| \frac{1}{2}, -\frac{1}{2} \right\rangle = (0 \ 1)\begin{pmatrix} 0 \\ 1 \end{pmatrix} = 1,$$
$$\left\langle \frac{1}{2}, \frac{1}{2} \middle| \frac{1}{2}, -\frac{1}{2} \right\rangle = \left\langle \frac{1}{2}, -\frac{1}{2} \middle| \frac{1}{2}, \frac{1}{2} \right\rangle = 0. \tag{36}$$

$$\left\langle \frac{1}{2}, \frac{1}{2} \middle| \frac{1}{2}, \frac{1}{2} \right\rangle = (1 \ 0)\begin{pmatrix} 1 \\ 0 \end{pmatrix} = 1,$$
$$\left\langle \frac{1}{2}, -\frac{1}{2} \middle| \frac{1}{2}, -\frac{1}{2} \right\rangle = (0 \ 1)\begin{pmatrix} 0 \\ 1 \end{pmatrix} = 1,$$
$$\left\langle \frac{1}{2}, \frac{1}{2} \middle| \frac{1}{2}, -\frac{1}{2} \right\rangle = \left\langle \frac{1}{2}, -\frac{1}{2} \middle| \frac{1}{2}, \frac{1}{2} \right\rangle = 0. \tag{37}$$

$$\left\langle \frac{1}{2}, \frac{1}{2} \middle| \frac{1}{2}, \frac{1}{2} \right\rangle = (1 \ 0)\begin{pmatrix} 1 \\ 0 \end{pmatrix} = 1,$$
$$\left\langle \frac{1}{2}, -\frac{1}{2} \middle| \frac{1}{2}, -\frac{1}{2} \right\rangle = (0 \ 1)\begin{pmatrix} 0 \\ 1 \end{pmatrix} = 1,$$
$$\left\langle \frac{1}{2}, \frac{1}{2} \middle| \frac{1}{2}, -\frac{1}{2} \right\rangle = \left\langle \frac{1}{2}, -\frac{1}{2} \middle| \frac{1}{2}, \frac{1}{2} \right\rangle = 0. \tag{38}$$

Now let us find and eigenvectors. First, observe that neither and have basis vectors as eigenvectors; their eigenvectors, however, can be written as follows in terms of .

$$| \psi_x\rangle_\pm = \frac{1}{\sqrt{2}}\left[\left|\frac{1}{2}, \frac{1}{2}\right\rangle \pm \left|\frac{1}{2}, -\frac{1}{2}\right\rangle\right],$$
$$| \psi_y\rangle_\pm = \frac{1}{\sqrt{2}}\left[\left|\frac{1}{2}, \frac{1}{2}\right\rangle \pm i\left|\frac{1}{2}, -\frac{1}{2}\right\rangle\right].$$

Therefore, and eigenvalue equations are provided by;

It is useful to introduce the Pauli matrices ,, when . These matrices are connected to the spin vector in the following ways (Chanyal, 2015).

$$\hat{\vec{S}} = \frac{\hbar}{2}\vec{\sigma}. \tag{39}$$

By applying this relation to Eqns. (30) and Eqn. (32), we obtain

$$\sigma_x = \begin{pmatrix} 0 & 1 \\ 1 & 0 \end{pmatrix}, \quad \sigma_y = \begin{pmatrix} 0 & -i \\ i & 0 \end{pmatrix}, \quad \sigma_z = \begin{pmatrix} 1 & 0 \\ 0 & -1 \end{pmatrix} \tag{40}$$

These matrices fulfill the following two conditions:

$$\sigma_j^2 = \hat{I} \qquad (j = x, y, z),$$
$$\sigma_j\sigma_k + \sigma_k\sigma_j = 0 \qquad (j \neq k), \tag{41}$$

Where are denoted by the subscripts j and k, and is the 2 × 2 unit matrix. The anticommutation relation is equal to these two equations (Grinevich et al., 2010).

$$\{\sigma_j, \sigma_k\} = 2\hat{I}\delta_{j,k}. \tag{42}$$

The Pauli matrices' satisfaction of the commutation relations can be confirmed.

$$[\sigma_j, \sigma_k] = 2i\, \varepsilon_{jkl}\sigma_l, \tag{43}$$

Where the Levi-Civita tensor, or antisymmetric tensor, is denoted by .

$$\varepsilon_{jkl} = \begin{cases} 1 & \text{if } jkl \text{ is an even permutation of } x, y, z, \\ -1 & \text{if } jkl \text{ is an odd permutation of } x, y, z, \\ 0 & \text{if any two indices among } j, k, l \text{ are equal.} \end{cases} \tag{44}$$

We can condense the relations Eqns. (41), (42), and (43) into (Hiroshima & Lőrinczi, 2008).

$$\sigma_j\sigma_k = \delta_{j,k} + i\sum_l \varepsilon_{jkl}\sigma_l. \tag{45}$$

We can confirm with this relation that, given any two vectors ; and ; that commute with we have

$$(\vec{\sigma}\cdot\vec{A})(\vec{\sigma}\cdot\vec{B}) = (\vec{A}\cdot\vec{B})\hat{I} + i\vec{\sigma}\cdot(\vec{A}\times\vec{B}), \tag{46}$$

The unit matrix is denoted by The determinants of the Hermitian, traceless Pauli matrices are -1.

$$\sigma_j^\dagger = \sigma_j, \qquad \mathrm{Tr}(\sigma_j) = 0, \qquad \det(\sigma_j) = -1 \qquad (j = x,\ y,\ z). \tag{47}$$

With and the connection , we obtain (Bezák, 1974).

$$\sigma_x\sigma_y\sigma_z = i\hat{I}. \tag{48}$$

We can demonstrate that using the commutation relations Eqn. (43).

$$e^{i\alpha\sigma_j} = I\cos\alpha + i\sigma_j\sin\alpha \qquad (j = x, y, z), \tag{49}$$

where α is an arbitrary real constant and I is the unit matrix (Ioffe & Neelov, 2003).

ACTIVITY 5.1

Objective: To investigate and understand the concept of spin angular momentum in quantum mechanics through a hands-on activity.

Materials Needed:

- Spinners or tops (can be purchased or homemade)
- Paper
- Pen
- Access to a quiet, spacious area

Instructions:

a. Introduction:
   - Begin by introducing the concept of angular momentum in quantum mechanics. Explain that in addition to orbital angular momentum, particles such as electrons possess intrinsic angular momentum known as spin. Discuss the theoretical background and importance of spin in quantum theory.

b. Activity Setup:
   - Distribute spinners or tops to each participant or group. These will represent electrons with spin.

c. Exploration of Spin:
   - Ask participants to spin their tops/spinners and observe their behavior.
   - Have them note the direction and stability of the spin. Encourage them to experiment with different spinning techniques (e.g., spin direction, speed).

d. Discussion
   - Gather participants to discuss their observations. Ask questions such as:
   - How does the spin of the top/spinner relate to the concept of spin in quantum mechanics?
   - What similarities and differences do you notice between the behavior of the top/spinner and the theoretical predictions about electron spin?
   - How might spin affect the properties and behavior of electrons in atoms?
   - Guide the discussion to cover topics such as spin quantization, spin states, and the Pauli Exclusion Principle.

SUMMARY

- The chapter on Angular Momentum in quantum mechanics covers a range of essential concepts. The topic starts by discussing Orbital Angular Momentum, which involves the process of quantization and the operators related to the rotating movement of particles around fixed locations.
- The General Formalism of Angular Momentum establishes the algebraic framework and defines commutation relations that are crucial for comprehending its quantum mechanical features. The Matrix Representation of Angular Momentum offers a tangible mathematical representation that simplifies calculations and enables theoretical forecasts.
- Geometrical Representation enhances understanding by visually representing angular momentum vectors in three-dimensional space, facilitating intuitive comprehension. The subsequent chapter is on Spin Angular Momentum, an inherent quality of particles, supported by Empirical Evidence that confirms its presence and defining features.
- The General Theory of Spin applies these ideas to quantum systems, specifically Spin 1/2 particles, and utilizes Pauli Matrices to describe their quantum states. Collectively, these subjects provide a thorough examination of the theoretical foundations and real-world consequences of angular momentum in the field of quantum mechanics.

REVIEW QUESTIONS

1. Define Orbital Angular Momentum in the context of quantum mechanics. How it quantized and what is are its operators?
2. Discuss the significance of commutation relations in the General Formalism of Angular Momentum. What do these relations imply about the observables involved?
3. Explain the Matrix Representation of Angular Momentum. Provide an example of how matrices are used to represent angular momentum operators.
4. How angular momentum is geometrically represented in three-dimensional space? What insights does this representation offer about angular momentum vectors?
5. What is Spin Angular Momentum? How does it differ from Orbital Angular Momentum? Provide examples of particles with intrinsic spin.
6. Describe experimental evidence supporting the existence and properties of Spin Angular Momentum. How do experiments verify the theoretical predictions about spin?
7. Explore the General Theory of Spin. How does it extend the concept of spin across different quantum systems?
8. Explain the significance of Spin 1/2 particles in quantum mechanics. How are Pauli matrices utilized to describe the properties of spin-1/2 particles?

MULTIPLE CHOICE QUESTIONS

1. **Which of the following correctly describes Orbital Angular Momentum?**
   a. It is associated with the intrinsic spin of particles.
   b. It quantizes the rotational motion of particles around a fixed point.
   c. It is represented by Pauli matrices.
   d. It is not observable in experiments.
2. **The General Formalism of Angular Momentum in quantum mechanics includes:**
   a. Geometrical representations only.
   b. Experimental evidence only.
   c. Only matrix representations.
   d. Commutation relations and mathematical operators.
3. **Which mathematical tool is commonly used for the Matrix Representation of Angular Momentum?**
   a. Dirac notation
   b. Schrödinger equation
   c. Pauli matrices
   d. Eigenfunctions
4. **Spin Angular Momentum refers to:**
   a. The intrinsic angular momentum of particles.
   b. The motion of particles around a nucleus.
   c. The classical notion of angular momentum.
   d. The orbital path traced by particles.
5. **Experimental evidence supporting Spin Angular Momentum includes:**
   a. Observations of particles rotating around fixed points.
   b. Measurements confirming discrete quantum states.
   c. Theoretical derivations of commutation relations.
   d. Predictions based on classical mechanics.
6. **Spin 1/2 particles are characterized by:**
   a. Integer values of angular momentum.
   b. Half-integer values of angular momentum.
   c. Lack of angular momentum.
   d. Negative values of angular momentum.

Answers to Multiple Questions

1. (b); 2. (d); 3. (c); 4. (a); 5. (b); 6. (c)

REFERENCES

1. Andreev, P. A. (2015). Quantum kinetics of spinning neutral particles: General theory and spin wave dispersion. Physica A: Statistical Mechanics and its Applications, 432, 108–126.
2. Aquilanti, V., Bitencourt, A. C. P., da S. Ferreira, C., Marzuoli, A., & Ragni, M. (2009). Combinatorics of angular momentum recoupling theory: Spin networks, their asymptotic, and applications. Theoretical Chemistry Accounts, 123(3), 237–247.
3. Asenjo, F. A., Zamanian, J., Marklund, M., Brodin, G., & Johansson, P. (2012). Semi-relativistic effects in spin-1/2 quantum plasmas. New Journal of Physics, 14(7), 073042.
4. Atkins, P. W., & Dobson, J. C. (1971). Angular momentum coherent states. Proceedings of the Royal Society of London. A. Mathematical and Physical Sciences, 321(1546), 321–340.
5. Atxitia, U., Chubykalo-Fesenko, O., Walowski, J., Mann, A., & Münzenberg, M. (2010). Evidence for thermal mechanisms in laser-induced femtosecond spin dynamics. Physical Review B—Condensed Matter and Materials Physics, 81(17), 174401.
6. Barke, I., Zheng, F., Rügheimer, T. K., & Himpsel, F. J. (2006). Experimental evidence for spin-split bands in a one-dimensional chain structure. Physical Review Letters, 97(22), 226405.
7. Benner, H. (1978). Experimental evidence for spin diffusion in the quasi-two-dimensional Heisenberg paramagnet $(C_2H_5NH_3)_2MnCl_4$. Physical Review B, 18(1), 319.
8. Bezák, V. (1974). A path-integral formulation of the Pauli equation using two real spin coordinates. International Journal of Theoretical Physics, 11, 321–339.
9. Bliokh, K. Y., Bekshaev, A. Y., & Nori, F. (2017). Optical momentum, spin, and angular momentum in dispersive media. Physical Review Letters, 119(7), 073901.
10. Chanyal, B. C. (2015). Octonion generalization of Pauli and Dirac matrices. International Journal of Geometric Methods in Modern Physics, 12(01), 1550007.
11. Chen, R., Li, N., Qian, H., Zhao, R. H., & Zhang, S. H. (2020). Experimental evidence refuting the assumption of phosphorus-31 nuclear-spin entanglement-mediated consciousness. Journal of Integrative Neuroscience, 19(4), 595–600.
12. Chen, R., Zhou, H., Moretti, M., Wang, X., & Li, J. (2019). Orbital angular momentum waves: Generation, detection, and emerging applications. IEEE Communications Surveys & Tutorials, 22(2), 840–868.
13. Dixon, W. G. (1970). Dynamics of extended bodies in general relativity. I. Momentum and angular momentum. Proceedings of the Royal Society of London. A. Mathematical and Physical Sciences, 314(1519), 499–527.
14. Dyall, K. G. (1994). An exact separation of the spin-free and spin-dependent terms of the Dirac-Coulomb-Breit Hamiltonian. The Journal of Chemical Physics, 100(3), 2118–2127.
15. Emmerich, K., Lippelt, E., Neuhaus, R., Pinkvos, H., Schwink, C., Gygax, F. N., & van der Wal, A. J. (1985). Experimental evidence for spatial inhomogeneous spin freezing in CuMn. Physical Review B, 31(11), 7226.

16. Florkowski, W., Friman, B., Jaiswal, A., Ryblewski, R., & Speranza, E. (2018). Spin-dependent distribution functions for relativistic hydrodynamics of spin-½ particles. Physical Review D, 97(11), 116017.

17. Franke-Arnold, S., Barnett, S. M., Yao, E., Leach, J., Courtial, J., & Padgett, M. (2004). Uncertainty principle for angular position and angular momentum. New Journal of Physics, 6(1), 103.

18. Gerratt, J. (1971). General theory of spin-coupled wave functions for atoms and molecules. In Advances in Atomic and Molecular Physics (Vol. 7, pp. 141–221).

19. Grinevich, P. G., Mironov, A. E. E., & Novikov, S. P. (2010). Zero level of a purely magnetic two-dimensional nonrelativistic Pauli operator for spin-1/2 particles. Theoretical and Mathematical Physics, 164, 1110–1127.

20. Hehl, F. W., & McCrea, J. D. (1986). Bianchi identities and the automatic conservation of energy-momentum and angular momentum in general-relativistic field theories. Foundations of Physics, 16(3), 267–293.

21. Herr, H., & Popovic, M. (2008). Angular momentum in human walking. Journal of Experimental Biology, 211(4), 467–481.

22. Hiroshima, F., & Lőrinczi, J. (2008). Functional integral representations of the Pauli–Fierz model with spin 1/2. Journal of Functional Analysis, 254(8), 2127–2185.

23. Hsiang, W. Y., & Straume, E. (2008). Global geometry of 3-body motions with vanishing angular momentum I. Chinese Annals of Mathematics, Series B, 29, 1–54.

24. Ioffe, M. V., & Neelov, A. I. (2003). Pauli equation and the method of supersymmetric factorization. Journal of Physics A: Mathematical and General, 36(10), 2493.

25. Keetels, G. H., Clercx, H. J. H., & van Heijst, G. J. F. (2008). Spontaneous angular momentum generation of two-dimensional fluid flow in an elliptic geometry. Physical Review E—Statistical, Nonlinear, and Soft Matter Physics, 78(3), 036301.

26. Leader, E., & Lorcé, C. (2014). The angular momentum controversy: What's it all about and does it matter? Physics Reports, 541(3), 163–248.

27. Leibfried, D., & Wineland, D. J. (2018). Efficient eigenvalue determination for arbitrary Pauli products based on generalized spin-spin interactions. Journal of Modern Optics, 65(5–6), 774–779.

28. Leyder, C., Romanelli, M., Karr, J. P., Giacobino, E., Liew, T. C., Glazov, M. M., & Bramati, A. (2007). Observation of the optical spin Hall effect. Nature Physics, 3(9), 628–631.

29. Light, G. L. (2013). The spinning motions of all fermions and bosons as implied by Pauli matrices containing complex conjugates in a combined spacetime four-manifold. Applied Physics Research, 5(4), 37.

30. Maeno, Y., NishiZaki, S., & Mao, Z. Q. (1999). Experimental evidence for spin-triplet superconductivity in Sr_2RuO_4. Journal of Superconductivity, 12, 535–541.

31. Mayer, M. G. (1950). Nuclear configurations in the spin-orbit coupling model. I. Empirical evidence. Physical Review, 78(1), 16.

32. Miao, P., Zhang, Z., Sun, J., Walasik, W., Longhi, S., Litchinitser, N. M., & Feng, L. (2016). Orbital angular momentum micro laser. Science, 353(6298), 464–467.

33. Moore, D. C., & Fleming, G. T. (2006). Angular momentum on the lattice: The case of nonzero linear momentum. Physical Review D—Particles, Fields, Gravitation, and Cosmology, 73(1), 014504.

34. Moses, H. E. (1970). Helicity representations of the coordinate, momentum, and angular momentum operators. Annals of Physics, 60(2), 275–320.

35. Mydosh, J. A. (2015). Spin glasses: Redux: An updated experimental/materials survey. Reports on Progress in Physics, 78(5), 052501.

36. Padgett, M., Courtial, J., & Allen, L. (2004). Light's orbital angular momentum. Physics Today, 57(5), 35–40.

37. Pan, W., Stormer, H. L., Tsui, D. C., Pfeiffer, L. N., Baldwin, K. W., & West, K. W. (2001). Experimental evidence for a spin-polarized ground state in the □=5/2 fractional quantum Hall effect. Solid State Communications, 119(12), 641–645.

38. Penrose, R. (1982). Quasi-local mass and angular momentum in general relativity. Proceedings of the Royal Society of London. A. Mathematical and Physical Sciences, 381(1780), 53–63.

39. Rakitzis, T. P., & Zare, R. N. (1999). Photofragment angular momentum distributions in the molecular frame: Determination and interpretation. The Journal of Chemical Physics, 110(7), 3341–3350.

40. Vaara, J., Jokisaari, J., Wasylishen, R. E., & Bryce, D. L. (2002). Spin–spin coupling tensors as determined by experiment and computational chemistry. Progress in Nuclear Magnetic Resonance Spectroscopy, 41(3–4), 233–304.

41. Wang, Y. L., Du, L., Xu, C. T., Liu, X. J., & Zong, H. S. (2014). Pauli equation for a charged spin particle on a curved surface in an electric and magnetic field. Physical Review A, 90(4), 042117.

42. Yao, A. M., & Padgett, M. J. (2011). Orbital angular momentum: Origins, behavior and applications. Advances in Optics and Photonics, 3(2), 161–204.

43. Zamanian, J., Marklund, M., & Brodin, G. (2010). Scalar quantum kinetic theory for spin-1/2 particles: Mean field theory. New Journal of Physics, 12(4), 043019.

CHAPTER 6

The Hydrogen Atom

LEARNING OBJECTIVES

After studying this chapter, you will be able to:

1. Understand classical and quantum descriptions of particles in central potentials.
2. Relate quantum numbers to energy levels and angular momentum properties.
3. Understand the energy levels and eigenfunctions of hydrogen atom states.
4. Relate quantum numbers to the spatial distribution of electron orbits.
5. Know the size and shape of hydrogen atom states using radial probability distributions.
6. Understand electric dipole transitions and their selection rules in hydrogen.
7. Understand the concept and implications of the reduced mass in hydrogen atom dynamics.
8. Explore the effects of relativistic corrections and spin-orbit interactions in hydrogen atom energy levels.

KEY TERMS FROM THIS CHAPTER

angular momentum
eigenfunctions and eigenvalues
quantum numbers
relativistic effects
selection rules
central potential
electric dipole transitions
reduced mass
Schrodinger equation
spin-orbit interaction

UNIT INTRODUCTION

The solar system functioned as the first significant trial devised to examine the ideas of classical mechanics. Conversely, the hydrogen atom played a crucial role as the first significant test to evaluate quantum mechanics. Both experiments were crucial for evaluating the theory of quantum mechanics. Each experiment holds a unique place in scientific history due to its significant implications (Bander & Itzykson, 1966).

The solar system was essential in validating Newtonian mechanics, as the predictable motion of planets around the sun demonstrated the accuracy and applicability of Newton's laws of motion and universal gravitation. This understanding was groundbreaking and laid the foundation for future advancements in physics. Observations and mathematical models of planetary orbits confirmed that classical mechanics could effectively describe large-scale celestial phenomena (Mills, 2000).

On the other hand, the hydrogen atom experiments marked a significant milestone in the development of quantum mechanics. The hydrogen atom has a nucleus carrying a positive charge of +e and an electron carrying a negative charge of -e. Upon first examination, the nucleus, which has a far larger mass than the electron, may seem to be a stationary object. The study of the hydrogen atom helped scientists understand the discrete energy levels and the wave-particle duality of electrons, challenging the classical view of particles (Dirac, 1926).

Given this, it should be possible to comprehend the characteristics of the hydrogen atom by solving a quantum mechanical problem that specifically examines a solitary particle, namely the electron, inside the Coulomb potential energy field (Stewart et al., 1965). The electron's behavior in the hydrogen atom, as dictated by the rules of quantum mechanics, demonstrated the quantization of energy levels and the probabilistic character of electron locations. This experiment not only validated the accuracy of quantum theory but also laid the foundation for the development of quantum mechanics as a strong and all-encompassing framework for comprehending atomic and subatomic phenomena (Parfitt & Portnoi, 2002).

$$V(r) = -\frac{e^2}{4\pi\epsilon_0 r}. \tag{1}$$

The objective of this chapter is to determine the energy eigenvalues and eigenfunctions of an electron, and then use these findings to explain the fundamental characteristics of the hydrogen atom. The next section will illustrate how this assessment might be enhanced by including the subtle influences resulting from relativity and the movement of the nucleus (Lamb & Retherford, 1951).

6.1. CENTRAL POTENTIALS

Examine the primary issue of a particle inside a central potential, such as the Coulomb potential, where the potential depends solely on the particle's distance from a fixed point. This scenario is fundamental in understanding various physical systems, including atomic structures and gravitational fields (Aashamar et al., 1978). The central potential, a function dependent on the radial distance from the center point, simplifies the issue by reducing it to one that just concerns the radial coordinate and excludes the angle coordinates. Streamlining this process is essential for resolving the Schrödinger equation in the science of quantum mechanics and for ascertaining the permissible energy levels and their related wavefunctions of the particle (Hendi et al., 2006).

6.1.1. Classical Mechanics of a Particle in a Central Potential

Analyze a classical particle characterized by a mass denoted as m. The particle has a vector position, denoted as r, and momentum, denoted as p. Furthermore, it exhibits an orbital angular momentum, denoted as L, which is precisely defined as the vector product of r and p with respect to the origin (Collas, 1981). If an object is moving inside a central potential V(r), it will encounter a force that may be mathematically represented as (Figure 6.1):

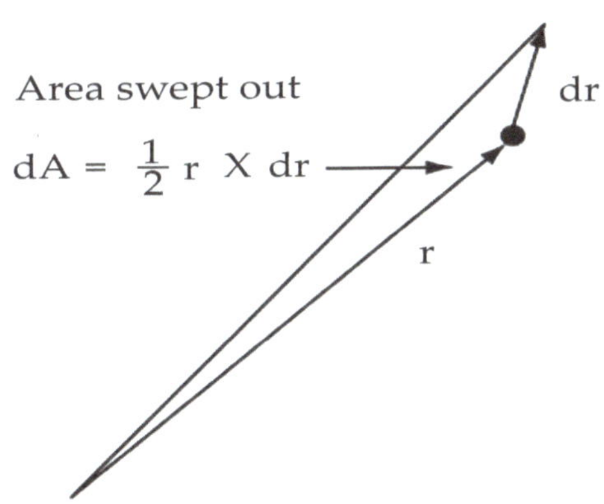

Figure 6.1. The radius vector r of a particle with angular momentum sweeps out a vector area in time .

Source: AC Phillip, Creative Commons License.

$$\mathbf{F} = -\frac{dV}{dr}\mathbf{e}_r$$

The symbol e_r denotes a unit vector that signifies the orientation of vector r. The torque exerted on the particle, as expressed by the equation N = r × F, is null since the force is applied in the direction of the radius vector r. Hence, the particle has a persistent rotating momentum symbolized with the symbol L (Nishiyama, 1965). The diagram demonstrates that the area bounded by the radius vector r during a small time period dt is governed by the equation:

$$d\mathbf{A} = \frac{\mathbf{L}}{2m} dt.$$

When the value of L stays constant, the particle's motion is limited to a certain plane. The particle's radius vector sweeps out the area at a constant pace of L/2m.

Momentum may be categorized into two distinct components: the radial component, which refers to motion along the radius, and the transverse component, which refers to motion perpendicular to the radius (Paradoksov, 1966). The components are mutually exclusive and may be easily taken into account while examining the motion of the particle:

$$p_r = m\frac{dr}{dt} \quad \text{and} \quad p_t = \frac{L}{m}.$$

The expression for total energy is as follows:

$$E = \frac{p_r^2}{2m} + \frac{L^2}{2mr^2} + V(r). \tag{2}$$

The energy of the particle may be seen as a combination of two separate elements: the radial kinetic energy, which is denoted by $p_r^2/2m$, and a viable potential energy expressed in a specific manner (Simon, 1954).

$$V_e(r) = \frac{L^2}{2mr^2} + V(r). \tag{3}$$

The effective force associated with this effective potential works radially and has a certain magnitude:

$$F_e = -\frac{dV_e}{dr} = \frac{L^2}{mr^3} - \frac{dV}{dr}.$$

The centrifugal force exerted outward may be mathematically expressed as L^2/mr^3, which is equivalent to mv^2/r for a particle traveling at a speed u along a circular route with a radius r as well as angular momentum L=mrv.

During its orbit around the sun, a celestial body with a mass indicated by the symbol m demonstrates gravitational potential energy:

$$V(r) = -\frac{GmM_{\odot}}{r}$$

Mø represents the mass of the sun, while G represents Newton's basic constant of gravity. If a planet were able to dissipate surplus energy, its orbit would ultimately become round. If the object were to gain sufficient energy to escape the gravitational pull of the sun, its orbit would transition from an ellipse to a parabola. Furthermore, if the object were to collect much more energy, its orbit would then become a hyperbola (Trushechkin & Volovich, 2009).

The rules of classical mechanics were initially tested in the context of planetary motion, an extra-terrestrial phenomenon that provided a natural laboratory for examining the validity of these laws. The success of classical mechanics in this domain was largely due to the fact that a planet's orbital angular momentum far exceeds the basic quantum unit of angular momentum, denoted as $\hbar$ (h-bar). This large angular momentum allowed classical mechanics to accurately predict planetary orbits without the need for quantum corrections.

To further exemplify the concept of motion in a central potential, focus on the hydrogen atom. In the context of the Coulomb potential, as described in Eqn. (1), an electron, carrying a negative charge of -e, revolves around a nucleus, which has a positive charge of +e. If the electron's angular momentum is far larger than $\hbar$, classical mechanics may be used to explain its movement. Under such circumstances, the electron traces a path that takes the shape of a conic section, like the orbits of planets revolving around the sun (Kryukov, 2013).

But with an electron's angular momentum on the order of $\hbar$, quantum mechanics replaces classical mechanics as the appropriate theory. Quantum mechanics is required in these situations to accurately describe the electron's behavior. Here, the electron is not simply following a well-defined orbit; instead, it occupies a quantum state characterized by a set of probabilities and wavefunctions. This quantum state possesses attributes that are fundamentally different from those predicted by classical mechanics (Muralidhar, 2021).

The distinction between classical and quantum mechanics becomes evident in the way they handle angular momentum. While classical mechanics allows for a continuous range of angular momentum values, quantum mechanics imposes quantization, meaning angular momentum can only take on discrete values that are integral multiples of $\hbar$. This leads to the formation of quantized energy levels and orbitals within the atom, a concept that classical mechanics cannot account for (Holas & March, 1990).

6.1.2. Quantum Mechanics of a Particle in a Central Potential

The attention will be directed towards quantum states characterized by precisely determined energy E, which possess wave functions that may be expressed in the following form:

$$\Psi(r,\theta,\phi,t) = \psi(r,\theta,\phi)\,e^{-iEt/\hbar}, \tag{4}$$

or,

$$\left[-\frac{\hbar^2}{2m}\nabla^2 + V(r)\right]\psi = E\psi. \tag{5}$$

To simplify this partial differential equation, one can make the assumption that the quantum state possesses both definite energy E and definite angular momentum characteristics. Given that the angular momentum of the orbit is represented by the equation and , the eigenfunctions may be calculated using the following method:

$$\psi(r,\theta,\phi) = R(r)Y_{l,m_l}(\theta,\phi). \tag{6}$$

Substitute Eqn. (6) into Eqn. (5),

$$\nabla^2\psi = \frac{1}{r}\frac{\partial^2(r\psi)}{\partial r^2} + \frac{1}{r^2}\left[\frac{\partial^2\psi}{\partial\theta^2} + \frac{\cos\theta}{\sin\theta}\frac{\partial\psi}{\partial\theta} + \frac{1}{\sin^2\theta}\frac{\partial^2\psi}{\partial\phi^2}\right]$$

And;

$$\hat{L}^2 = -\hbar^2\left(\frac{\partial^2}{\partial\theta^2} + \frac{\cos\theta}{\sin\theta}\frac{\partial}{\partial\theta} + \frac{1}{\sin^2\theta}\frac{\partial^2}{\partial\phi^2}\right),$$

$$-\frac{\hbar^2}{2mr}\frac{d^2(rR)}{dr^2} + \left[\frac{l(l+1)\hbar^2}{2mr^2} + V(r)\right]R = ER. \tag{7}$$

By introducing u(r);

$$R(r) \equiv \frac{u(r)}{r}, \tag{8}$$

Or;

$$-\frac{\hbar^2}{2m}\frac{d^2u}{dr^2} + \left[\frac{l(l+1)\hbar^2}{2mr^2} + V(r)\right]u = Eu. \tag{9}$$

The equation derived is known as the radial Schrödinger equation, which holds significant importance in the study of quantum mechanics. This equation is particularly relevant when examining a particle that possesses angular momentum. In such scenarios, the particle exhibits behavior analogous to that of a particle confined to a one-dimensional (1-D) effective potential, characterized by a specific mathematical form (Bromberg, 1976).

The radial Schrödinger equation is obtained by using the spherical symmetry of the issue, derived from the whole three-dimensional Schrödinger equation. By decomposing the factors into radial as well as angular components, the calculation simplifies to a more tractable form that characterizes the radial aspect of the wave function. By reducing, it becomes a one-dimensional equation that only involves the radial coordinate (Stevenson, 1984).

$$V_e(r) = \frac{l(l+1)\hbar^2}{2mr^2} + V(r). \tag{10}$$

The boundary condition is:

$$u(r) = 0 \quad \text{at} \quad r = 0$$

To ensure that the function R(r) = u(r)/r remains finite, and therefore the three-dimensional eigenfunction stated in Eqn. (6) is valid, certain conditions must be imposed at the origin. Furthermore, the boundary condition must be satisfied by bound state solutions (Voth, 1996), which refer to particles that are unable to escape to infinity:

$$u(r) \to 0 \quad \text{as} \quad r \to \infty.$$

Bound states are present only when the effective potential $V_e(r)$, as defined in Eqn. (10), exhibits a significant level of attraction. The states will be designated by a quantum number n_r, which takes on values of 0, 1, 2, and so on (Mielnik, 1974). The eigenfunction can be represented by a specific form:

$$\psi_{n_r,l,m_l}(r,\theta,\phi) = \frac{u_{n_r,l}(r)}{r}\, Y_{l,m_l}(\theta,\phi). \tag{11}$$

By applying the normalization criterion to spherical harmonics, the eigenfunction is considered to be normalized if the radial eigenfunction satisfies the requirement (Rizov et al., 1985).

$$\int_0^\infty |u_{n_r,l}|^2 \, dr = 1. \tag{12}$$

Use the function to get the energy of bound states. The energy is directly proportional to the variable "l" because it determines the average transverse kinetic energy. Furthermore, the energy is directly correlated with the variable "nu" since the average radial kinetic energy, which is determined by it, increases with the number of radial nodes (Jelic & Marsiglio, 2012).

$$\int_0^\infty u^*_{n_r,l}(r) \left(\frac{l(l+1)\hbar^2}{2mr^2} \right) u_{n_r,l}(r) \, dr;$$

Before delving into the precise equations for the energy levels and eigenfunctions of a particle in a Coulomb potential, first revisit the methodology used to construct the radial Schrodinger equation, as shown in Eqn. (9) (Barton, 1986).

In addition to this, a particle that is in a central potential has a quantum state that is characterized by a certain energy as well as another observable attribute that has a precise value. The term used to describe this is "parity" (Condon & Morse, 1931). Eigenfunctions possessing the specified attribute:

$$\psi(-\mathbf{r}) = +\psi(\mathbf{r}), \tag{13}$$

Have even parity, and eigenfunctions

$$\psi(-\mathbf{r}) = -\psi(\mathbf{r}) \tag{14}$$

Have odd parity.

6.2. QUANTUM MECHANICS OF THE HYDROGEN ATOM

Through the use of quantum mechanical principles, a summary of the hydrogen atom will be presented in this subsequent section. The hydrogen atom, being one of the simplest atomic systems, consists of a single electron orbiting a nucleus under the influence of a Coulomb potential. The electrical attraction between the negatively charged electron and the positively charged proton, or nucleus, produces the Coulomb potential (Dirac, 1926).

Given this setup, it is reasonable to infer that the bound states of atomic hydrogen possess well-defined orbital angular momentum characteristics. The solutions to the hydrogen atom Schrödinger equation, covered in the previous section, determine the attributes of these bound states (Pauling, 1928).

The Schrödinger equation for the hydrogen atom is solved by separating the variables into radial and angular components. The functions specified by the quantum numbers ⊠ and ⊠, which denote the angular momentum of the orbit and its projection along a selected axis, respectively, explain the angular component of the wavefunction. These functions are called spherical harmonics. These quantum numbers determine the shape and orientation of the electron's probability distribution, or orbital (Casati et al., 1987). Thus:

$$\Psi_{n_r,l,m_l}(r,\theta,\phi,t) = \psi_{n_r,l,m_l}(r,\theta,\phi)\,\exp(-iE_{n_r,l}t/\hbar)$$

With

$$\psi_{n_r,l,m_l}(r,\theta,\phi) = \frac{u_{n_r,l}(r)}{r}\,Y_{l,m_l}(\theta,\phi),$$

One of the solutions to the problem is the radial eigenfunction

$$\frac{\hbar^2}{2m_e}\frac{d^2u_{n_r,l}}{dr^2}+\left[\frac{l(l+1)\hbar^2}{2m_e r^2}-\frac{e^2}{4\pi\epsilon_0 r}\right]u_{n_r,l}=E_{n_r,l}u_{n_r,l},\tag{15}$$

With boundary conditions:

$$u_{n_r,l}(r)=0 \quad \text{at} \quad r=0 \quad \text{and at} \quad r=\infty.\tag{16}$$

The properties of the energy levels arising from the eigenvalue issue determined by Eqns. (15) and (16) may be deduced by analyzing the effective potential in Eqn. (15).

$$V_e(r)=\frac{l(l+1)\hbar^2}{2m_e r^2}-\frac{e^2}{4\pi\epsilon_0 r}.\tag{17}$$

Figure 6.2 displays the potential form for electrons with varying values of the quantum number, l, which represents orbital angular momentum. For values of l that are not zero, it has been observed that the effective potential is attractive when r is big and repulsive when r is small (Campos et al., 2018). When the derivative of is set to zero, the function (r) reaches its minimum value:

$$V_e(r)=-\frac{E_R}{l(l+1)} \quad \text{at} \quad r=l(l+1)a_0,\tag{18}$$

The Bohr radius is:

$$a_0=\left[\frac{4\pi\epsilon_0}{e^2}\right]\frac{\hbar^2}{m_e}=0.529\times10^{-10}\ \text{m}\tag{19}$$

And the Rydberg energy is:

$$E_R=\frac{e^2}{8\pi\epsilon_0 a_0}=13.6\ \text{eV}.\tag{20}$$

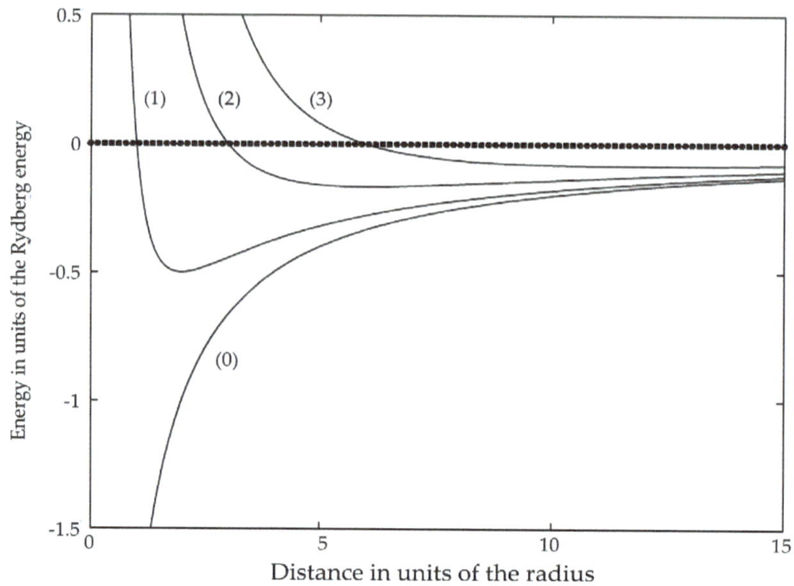

Figure 6.2. Illustration of the effective potential energy for an electron in a hydrogen atom with orbital angular momentum quantum numbers l = 0, 1, 2, and 3.

Source: Feynman, R.P, Creative Commons License.

Figure 6.2 illustrates that the effective potential experienced by an electron with zero angular momentum is exclusively attracting. In the context of classical mechanics, an electron moves rapidly towards a proton without any stable bound states (Gordeyev & Chhajlany, 1997).

As per the uncertainty principle, when an electron is confined inside a region of size u, its momentum becomes uncertain and is approximately on the scale of . Additionally, the average kinetic energy of the electron is at least on the order of.

Therefore;

$$E \approx \frac{\hbar^2}{2m_e r^2} - \frac{e^2}{4\pi\epsilon_0 r}.$$

As the region of localization diminishes, the energy drops due to a decrease in potential energy. However, eventually, the kinetic energy of localization grows at a faster rate. The total energy achieves the lowest threshold of approximately (Mills, 2005):

$$E \approx -E_R \quad \text{at} \quad r \approx a_0.$$

The minimal value is an approximation of the lowest achievable energy of an electron with no angular momentum in a Coulomb potential (Cisneros & McIntosh, 1969). It also implies the existence of quantum-bound states with energies falling within a specific range: and E = 0.

6.2.1. Energy Levels and Eigenfunctions

It can be demonstrated that an electron with angular momentum L in a Coulomb potential possesses an unlimited number of bound states, and their energies may be determined using the following equation (Mirlin, 2000):

$$E_{n_r,l} = -\frac{E_R}{(n_r + l + 1)^2} \quad \text{with} \quad n_r = 0, 1, 2, 3, \ldots \tag{21}$$

As anticipated, there are bound states that possess both zero and non-zero angular momentum. Furthermore, in cases when the angular momentum is considerably elevated, the energy levels exhibit a dense arrangement, indicating a correlation with a continuous spectrum of classical energies linked to bound states (Huang & Carrington, 2000).

Surprisingly, a significant number of energy levels in Figure 6.3, which share the same value for (n_r+l), exhibit identical energy. Therefore:

$$E_n = -\frac{E_R}{n^2} \tag{22}$$

Where; is a quantum number:

$$n = n_r + l + 1. \tag{23}$$

Where; n = 1, 2, 3.

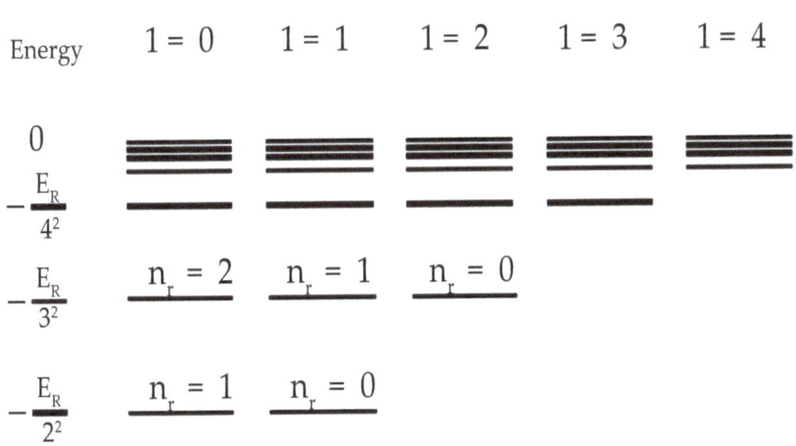

Figure 6.3. Illustration of the energy levels given by eqn. (21) for the bound states of an electron in a hydrogen atom.

Source: Townsend, J.S, Creative Commons License.

Did you know?

The energy levels of the hydrogen atom are discretized, indicating that electrons can only exist in distinct energy states. Quantization is a key notion in quantum physics that has significant consequences for comprehending atomic structure along with behavior. The quantization of energy levels implies that the electron cannot possess random energies, but rather it must inhabit one of the distinct energy states determined by the quantum number n.

Section 6.7 will demonstrate that the radial eigenfunction associated with the energy eigenvalue possesses three distinct features:

a. The radial eigenfunction of an electron in a Coulomb potential (Ching & Huber, 1982), where the electron must be confined to a certain amount of energy according to the Coulomb potential:

$$\epsilon = \frac{E_R}{n^2} = \frac{\hbar^2}{2m_e n^2 a_0^2}$$

Falls off exponentially at large r like:

$$u_{n_r,l}(r) \propto e^{-r/na_0}.$$

b. The behavior of the eigenfunction at small values of u is dictated by the orbital angular momentum, which is influenced by the distinct nature of the centrifugal potential at u = 0. This behavior can be described as (Gnanasangeetha, 2022):

$$u_{n_r,l}(r) \propto r^{l+1}.$$

c. The symbol "nu" represents the number of nodes located between. :

$$u_{n_r,l}(r) \propto p_{n_r,l}(r).$$

By combining these three characteristics, a radial eigenfunction results that has the following structure (Neuhauser, 1990):

$$u_{n_r,l}(r) = N p_{n_r,l}(r) r^{l+1} e^{-r/na_0}, \tag{24}$$

N is a constant that guarantees the fulfillment of the normalization requirement Eqn. (12).

Some of the eigenfunctions are depicted in Figure 6.4. The eigenfunctions are designated using spectroscopic notation in order to adhere to the standards of atomic physics (Çetin, 2017).

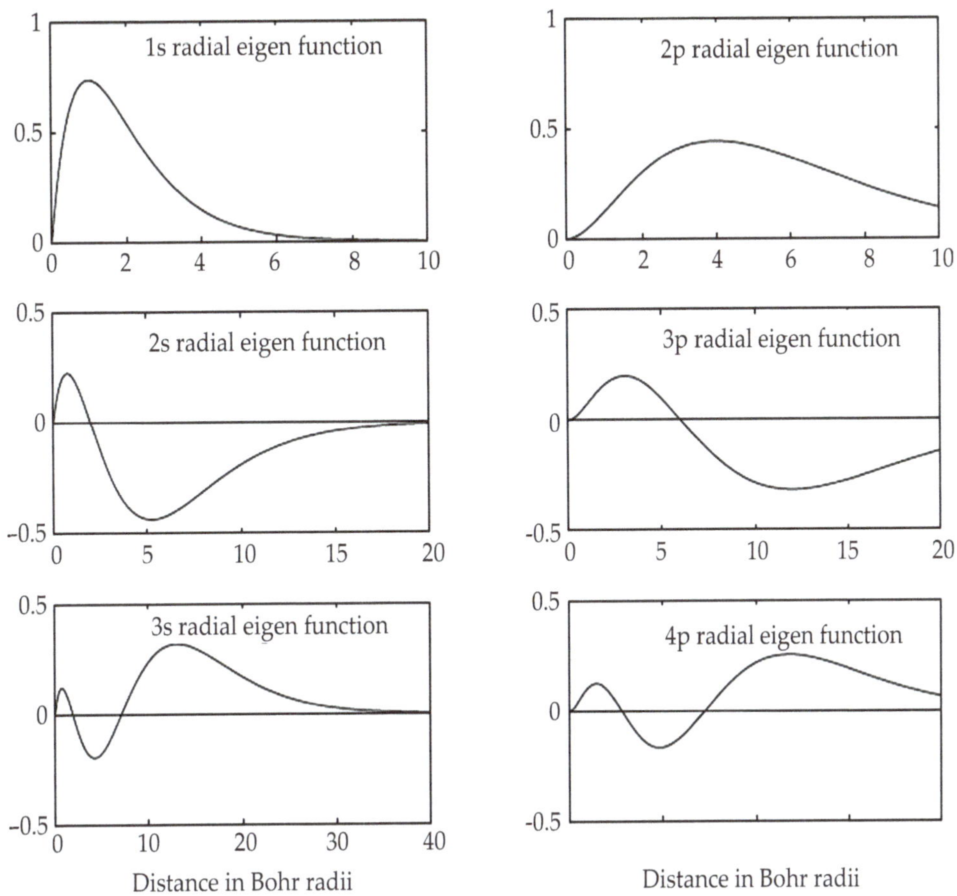

Figure 6.4. Illustration of Radial eigenfunctions for an electron in the hydrogen atom with radial quantum numbers = 0, 1, and 2 and with angular momentum quantum numbers l = 0 and 1.

Source: Park, D., Creative Commons License.

6.3. SIZES AND SHAPES

By examining the electron's regions of highest probability inside the atom, it is possible to ascertain the dimensions and configuration of a quantum state of atomic hydrogen (Frosini, 1992). The probability of locating the electron at the coordinates inside a volume element is high for a state characterized by the eigenfunction .

$$|\psi_{n_r,l,m_l}(r,\theta,\phi)|^2 \, d^3\mathbf{r}.$$

It is possible to ascertain the radial probability distribution. Utilize the expression , where represents a component of solid angle (Yamamoto et al., 2004). Demonstrate the eigenfunction as:

$$\psi_{n_r,l,m_l}(r,\theta,\phi) = \frac{u_{n_r,l}(r)}{r} Y_{l,m_l}(\theta,\phi),$$

The likelihood of locating the electron within a range of distances between r and r + dr from the nucleus can be expressed as:

$$\left|\frac{u_{n_r,l}(r)}{r}\right|^2 r^2 \, dr \int |Y_{l,m_l}(\theta,\phi)|^2 \, d\Omega = |u_{n_r,l}(r)|^2 \, dr.$$

Figure 6.5 displays the radial probability densities for states with primary quantum number n_r = 0, 1, and 2, along with azimuthal quantum number l = 0 and 1. It is observed that the radial extent expands with increasing values of n_r and l (Salzemann et al., 2004). The average radius is:

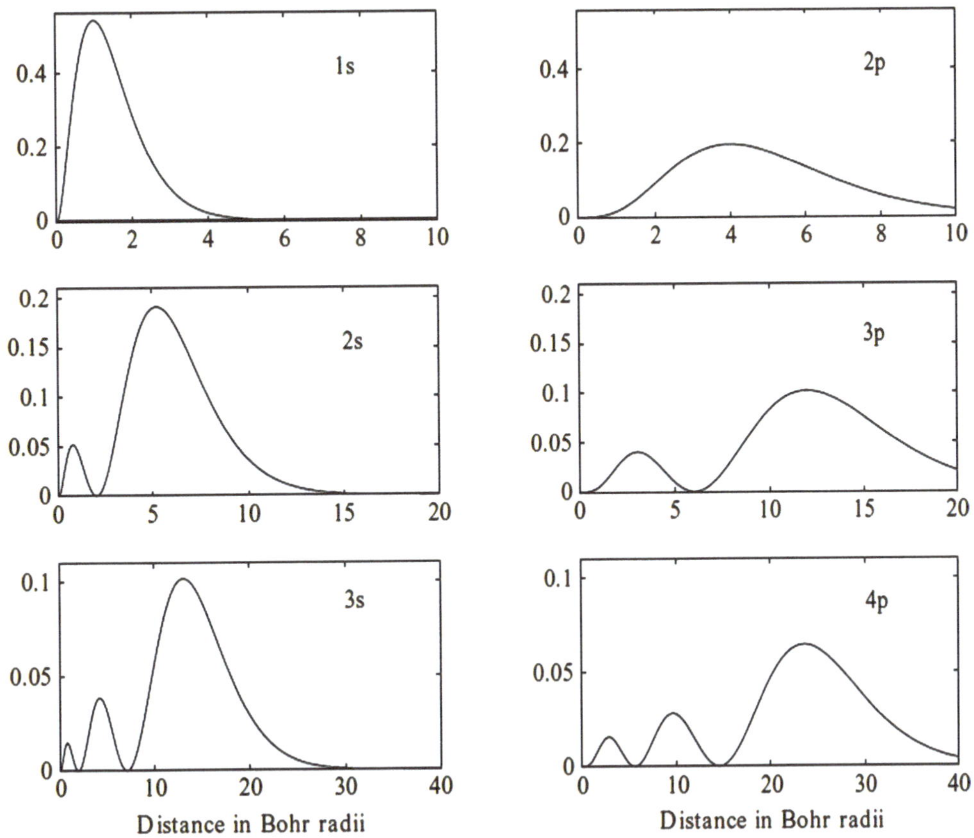

Figure 6.5. Radial probability densities for the 1s, 2s, 3s, 2p, 3p and 4p states of the hydrogen atom. Note that the unit of distance is the Bohr radius and that a different scale is used for states with a different number of radial nodes.

Source: Liboff, R.L, Creative Commons License.

$$\langle r \rangle_{n_r,l} = \int_0^\infty r|u_{n_r,l}(r)|^2 \, dr.$$

The integral may be represented by using the mathematical characteristics of the eigenfunctions.

$$\langle r \rangle_{n_r,l} = \tfrac{1}{2}[3n_r^2 + 6n_r(l+1) + (l+1)(2l+3)]a_0, \tag{25}$$

The probability density of the state stays invariant upon rotation around the z-axis. Thus, the state's size and arrangement may be accurately ascertained by exhibiting the electron's positions with the highest probability on any vertical plane (Kalendová, 2003) that intersects the z-axis, as shown in Figures 6.6 and 6.7.

Figures 6.6 and 6.7 depict the measurements and arrangements of the 3p and 3d states of atomic hydrogen. The states shown in Figure 6.6, which include three primary quantum numbers, exhibit a single node that extends radially. Additionally, these states have an angular dependency denoted as , where m_l may take the values of 1, 0, and -1. The states shown in Figure 6.7 possess a three-dimensional structure and do not exhibit any radial nodes (Marcus, 1994).

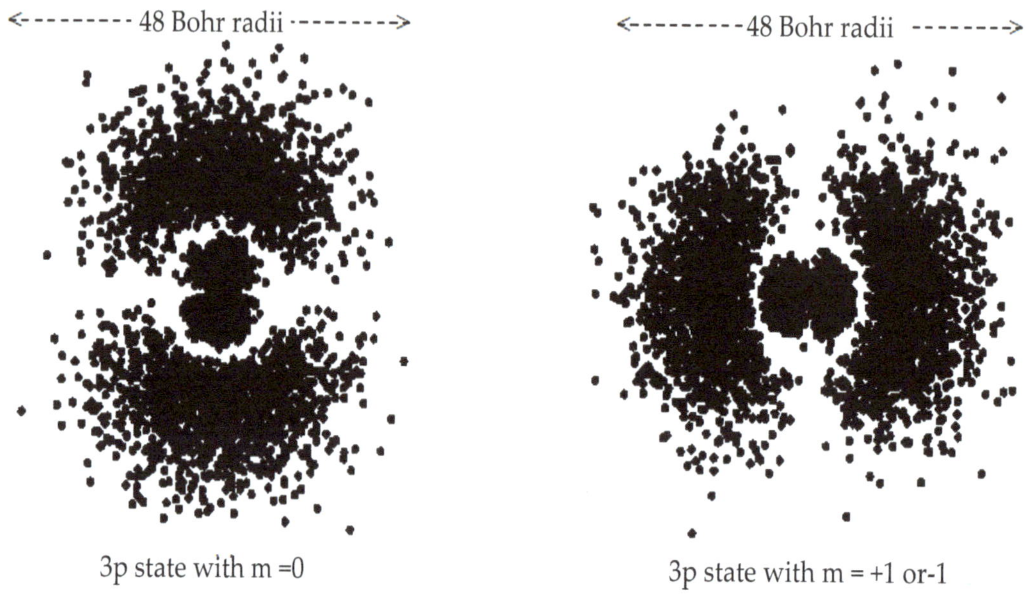

Figure 6.6. Illustration of the size and shape of the 3p states of the hydrogen atom with a z component of orbital angular momentum equal to . These states have rotational symmetry about the z axis. The density of dots is proportional to the probability of finding an electron on a vertical plane passing through the z axis. These pictures were produced by selecting a point on the plane at random and deciding to plot or not to plot in accordance with the value of at that point.

Source: Griffiths, D.J., Creative Commons License.

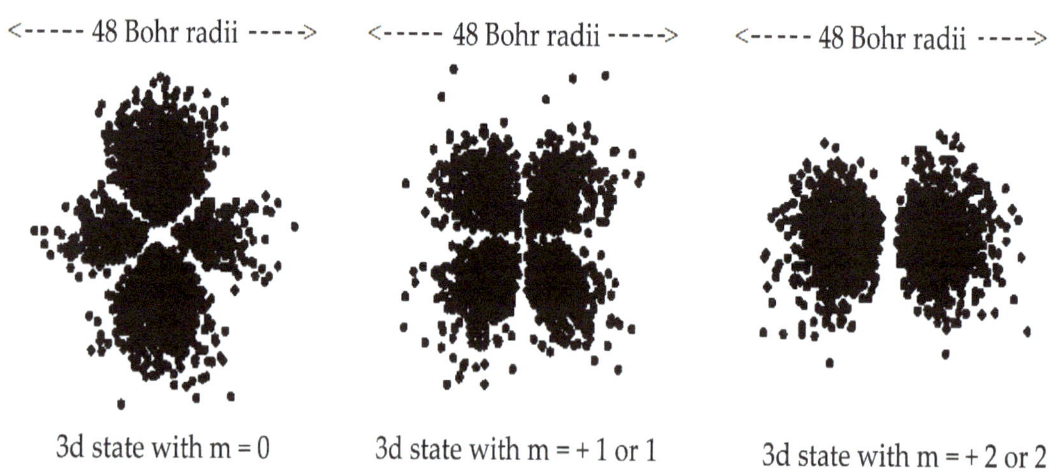

Figure 6.7. Illustration of the size and shape of the 3d states of the hydrogen atom with a z component of orbital angular momentum equal to .

Source: Mandl, F, Creative Commons License.

6.4. RADIATIVE TRANSITIONS

Transitions known as electric dipole transitions are the most common type of radiative transitions in atomic systems. These happen as a result of the interaction between the electromagnetic field's electric field component and the electron-nucleus system's electric dipole moment operator (Kubo & Toyozawa, 1955).

The electric dipole moment operator describes the distribution of electric charge within the atom and its response to an external electric field. This interaction leads to transitions between different energy levels in the atom.

Selection criteria dictate the permissible transitions in electric dipole moments, depending on the characteristics of the electric dipole moment operator and the atomic wavefunctions. These rules dictate, for example, the changes in orbital angular momentum dl and the magnetic quantum number m that are permissible during such transitions (Colangelo et al., 2005).

The interaction is represented by:

$$\hat{H}_I = -\mathbf{d} \cdot \mathbf{E}. \tag{26}$$

The existence of this interaction directly correlates with the likelihood of a transition happening between states classified by quantum numbers n_i, l_i, and ml_i, and states described by quantum numbers n_f, l_f, and ml_f.

$$\left| \int \psi^*_{n_f, l_f, m_{l_f}}(\mathbf{r})\ \hat{H}_I\ \psi_{n_i, l_i, m_{l_i}}(\mathbf{r})\ \mathrm{d}^3\mathbf{r} \right|^2. \tag{27}$$

By demonstrating that the integral in Eqn. (27) is equal to zero when the starting and end states have identical parities, one may easily provide proof to support the claim that electric dipole transitions always include a change in parity. This is shown by analyzing the outcomes that arise from replacing the integration variable r with -r during the whole course of the experiment. If the eigenfunction has even parity, the sign of the interaction H^I = —dE is reversed, but the sign of the eigenfunction remains the same. Conversely, the sign of the eigenfunction is altered when the eigenfunction exhibits odd parity, as seen in Eqns. (13) and (14). Therefore, if the eigenfunctions have identical parity, the integrand in Eqn. (27) will experience a sign reversal when the integration variable r is replaced with -r in the equation. This implies that the value of the integral must be precisely zero in order to be considered genuine (Huang & Rhys, 1950).

$$\Delta l = \pm 1. \tag{28}$$

The dot lines in Figure 6.8 represent the transitions between the low-lying states of the hydrogen atom, which are characterized by electric dipole interactions. The transitions are denoted using spectroscopic notation, such as 1s, 2s, 2p, and so on. Certain values of the primary quantum number n and orbital angular momentum quantum number l correspond to these designations. For example, the state with n = 2 and l = 0 is denoted by the symbol 2s. This refers to an electron located in the second energy level (or shell) of a hydrogen atom, having no orbital angular momentum (a spherical orbital shape). The state with n = 2 and l = 1 is represented by the symbol 2p, which is different from the previous state. Even though the electron is located in the second energy level, it has a quantum number of one for its orbital angular momentum, which corresponds to an orbital that is shaped like a dumbbell (Grant, 1974). These spectroscopic notations provide a concise way to identify and discuss different energy levels and orbital configurations within the hydrogen atom. They are crucial for understanding the transitions depicted in Figure 6.8, which show how electrons move between these distinct states, emitting or absorbing photons in the process (Aad et al., 2012).

$$\hbar w = |E_{nf} - E_{ni}|;$$

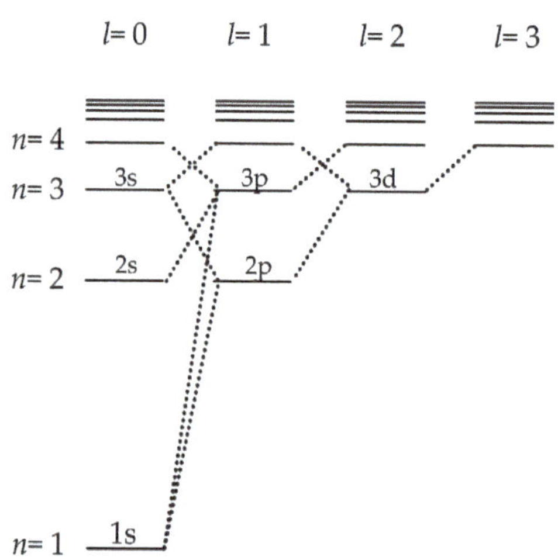

Figure 6.8. Electric dipole radiative transitions between low-lying energy levels of the hydrogen atom with different values for the quantum numbers n and l. Spectroscopic notation, 1s, 2s, 2p, etc. has been used to label the energy levels; for example, 2s corresponds to n 2 and l 0 and 2p corresponds to n 2 and l 1. We note that the electric dipole transitions shown by the dotted lines obey the Δl 1 selection rule given in Eqn. (28).

Source: Peebles, P.J.E, Creative Commons License.

There are two possible explanations for the transitions seen in Figure 6.8: either they are the product of an external influence or they occur spontaneously without any external force. Strong transitions are characterized by the presence of large amounts of energy:

Electromagnetic energy is emitted when the final energy state (En_f) exceeds the initial energy state (En_i), whereas electromagnetic energy is absorbed when the final energy state (En_f) exceeds the initial energy state (En_i). Spontaneous transitions, although they may seem to occur without any specific reason, are really the outcome of the atom's interacting with a quantized radiation field that is consistently present, even when the atom is completely isolated. An atom with an initial energy of En_i transitions to an atom with a final energy of En_f due to this interaction. This transition takes place by the emission of a photon with an energy equivalent to the disparity among the initial and final energies (En_i-En_f). These transitions lead to the creation of spectral lines (Krupke, 1971), which are defined by certain wavelengths denoted by the symbol λ:

$$\frac{hc}{\lambda} = E_R\left(\frac{1}{n_f^2} - \frac{1}{n_i^2}\right). \tag{29}$$

The mentioned lines constitute a sequence of lines in the ultraviolet region known as the Lyman series, characterized by specific wavelengths:

$$\frac{hc}{\lambda} = E_R\left(\frac{1}{1^2} - \frac{1}{n_i^2}\right), \quad \text{with} \quad n_i = 2, 3, 4 \ldots,$$

The Balmer series refers to a sequence of observable lines characterized by specific wavelengths:

$$\frac{hc}{\lambda} = E_R\left(\frac{1}{2^2} - \frac{1}{n_i^2}\right), \quad \text{with} \quad n_i = 3, 4, 5 \ldots,$$

As well as other series characterized by longer wavelengths.

Upon careful examination of Figure 6.8, it is evident that the 2s state of the hydrogen atom is immune to decay when exposed to electric dipole radiation. According to Eqn. (28), transitioning from the 2s state to the 1s state would go against the $\Delta l = 1$ rule. This is the cause behind this situation. The term "2s state" is used to describe a condition that is metastable and has a prolonged duration. Instead of undergoing electric dipole transfer, it ultimately transitions to the 1s state by a less probable method. The mean duration for an abrupt shift from the 2p state to the 1s state is $1.6 \times 10^{(-9)}$ seconds. Conversely, the typical time it takes for a spontaneous transition from the 2s condition to the 1s state is 0.14 seconds, which is considerably longer (Finkelstein et al., 2021).

6.5. THE REDUCED MASS EFFECT

Up until this point, the operation has been made under the premise that the mass of the nucleus of the hydrogen atom is sufficiently large to ignore its movement. This movement results in implications that are not immediately obvious but significant (MacMahon et al., 1986).

Through the use of a Coulomb potential, one may ascertain the classical energy of an electron by analyzing the interaction that occurs between the electron and another particle or particles.

$$E = \frac{p_e^2}{2m_e} + \frac{p_N^2}{2m_N} - \frac{e^2}{4\pi\epsilon_0 r},$$

In the context of the center-of-mass frame, one may define the magnitudes of the momenta of the electron and the nucleus as p_e and p_N, respectively. The distance between the two is represented by the symbol u. If $p_e = p_N = p$, one may derive a formula for the energy in the following manner (Kasiske et al., 1995):

$$E = \frac{p^2}{2\mu} - \frac{e^2}{4\pi\epsilon_0 r}, \quad \text{where} \quad \mu = \frac{m_e m_N}{m_e + m_N}.$$

When the traditional electron-nucleus system is seen in the center of the mass frame, it exhibits similar behavior to that of a single particle with a mass of around μ. Within the electron-nucleus system, the mass in question is known as the decreased mass of the system, a phrase used to describe it (Kram et al., 1997).

By the same token, a quantum electron-nucleus system exhibits the characteristics of a unified entity, characterized by a reduced mass μ. To be more specific, the Bohr radius a_0, which is specified in Eqn. (19), and the Rydberg energy Es_R, which is described in Eqn. (20), are subjected to minor adjustments and are changed into:

$$a_0' = \left[\frac{4\pi\epsilon_0}{e^2}\right]\frac{\hbar^2}{\mu} = \frac{m_e}{\mu}a_0$$

and

$$E'_R = \frac{e^2}{8\pi\epsilon_0 a'_0} = \frac{\mu}{m_e} E_R,$$

The quantitative results that were presented in Eqns. (21), (22), (24), and (25), as well as in Figures 6.3, 6.4, and 6.5, have been modified to reflect the new information. It is important to note that the energy levels of the hydrogen atom (Donelan & Kram, 1997) have been modified and are now stated as:

$$E'_n = -\frac{E'_R}{n^2} = \frac{\mu}{m_e} \frac{E_R}{n^2}.$$

(30)

The nucleus of a typical hydrogen atom is made up of a proton that has a mass that is 1836 times greater than the mass of an electron (1836m_e). This results in a reduced mass of (1836/1837)m_e. On the other hand (Kobrin et al., 1984), a heavy hydrogen atom has a deuteron nucleus that has a mass that is 3671 times greater than the mass of an electron (3671m_e), which results in a reduced mass of (3671/3672)m_e. Because of the tiny difference in the masses of these two particles, there is a noticeable variation in the wavelengths of the emitted spectral lines (Gu et al., 2003). As an example, the H_α line in the Balmer series corresponds to the change from a state with n = 3 to n = 2 and may be characterized by the wavelength λ:

$$\frac{hc}{\lambda} = \frac{\mu}{m_e} E_R \left[\frac{1}{2^2} - \frac{1}{3^2}\right].$$

The wavelength of regular hydrogen, measured at 656.4686 nm, is marginally greater than the wavelength of deuterium, measured at 656.2899 nm. In 1934, Harold Urey made a significant breakthrough by discovering heavy hydrogen through an experiment that showed the existence of a faint line alongside each line in the Balmer series. The presence of this weak line was caused by the amalgamation of a small amount of deuterium with naturally occurring hydrogen (Lass et al., 2018).

6.6. RELATIVISTIC EFFECTS

This section will demonstrate that the electron in a hydrogen atom can be regarded as non-relativistic, and will be done with a high degree of precision. On the other hand, it is essential to keep in mind that relativistic effects do, in fact, bring about alterations to the energy levels that are both quite tiny and quite large (Pyykkö, 2012)

Through the utilization of Eqn. (19), it has been determined that:

$$p_0 = \alpha m_e c,$$

(31)

Where; is called the fine structure constant:

$$\alpha = \frac{e^2}{4\pi\epsilon_0 \hbar c} = \frac{1}{137.035\,989\,5}.$$

(32)

In the field of atomic physics, the relatively low numerical value of the fine structure constant has a variety of important implications. It suggests that atomic electrons, such as the one found in a hydrogen atom, have momenta that are far less than $m_e c$, which indicates that non-relativistic physics is an appropriate approximation for the situation. To assess the extent of relativistic adjustments made to the hydrogen atom (Schwerdtfeger, 2002), one may examine the relationship between the momentum (p) and relativistic energy (c) of an electron.

$$\epsilon = \sqrt{m_e^2 c^4 + p^2 c^2}.$$

If p is much less than m_e, the binomial theorem can be employed to obtain:

$$\epsilon = m_e c^2 + \frac{p^2}{2m_e} - \frac{1}{8}\left(\frac{p}{m_e c}\right)^4 m_e c^2.$$

The first component represents the energy associated with the rest mass, the second component represents the kinetic energy in non-relativistic conditions, and the third component represents the main correction to the kinetic energy due to relativity (Helings, 1986). It is important to note that the average momentum of an electron in a hydrogen atom, approximately equal to $p_0 = \alpha m_e c$, results in a relativistic correction to its kinetic energy of similar magnitude:

$$\langle E_{rel} \rangle \approx -\frac{1}{8}\left(\frac{p_0}{m_e c}\right)^4 m_e c^2 = -\frac{1}{8}\alpha^4 m_e c^2. \tag{33}$$

Likewise, analogous corrections of equal size occur as a result of the interaction between an electron's spin magnetic moment and a magnetic field generated by the relative movement of a nucleus and an electron. The term used to describe the phenomenon under discussion is the interaction between spin and orbit (Van Lenthe et al., 1999).

The velocity of a classical electron that is circling the nucleus in a circular route with a radius of u can be analyzed to obtain an estimation of the amplitude of the spin-orbit interaction. When the electron has a velocity u and an orbital angular momentum L, the time is equal to the product of these two quantities:

$$\tau = \frac{2\pi r}{\upsilon} = \frac{2\pi m_e r^2}{L}$$

in order to complete a single revolution around a celestial body. Similar to the perception of the sun's movement around the earth, the electron observes a nucleus that seems to orbit in a circular path with a radius of u and a period of t. The electron forms a current loop due to the electric charge e of the nucleus (Dimopoulos et al., 2008), resulting in an electric current $I = e/\tau$. The magnetic field experienced is:

$$B = \frac{e}{4\pi\epsilon_0 m_e c^2 r^3} L. \tag{34}$$

The spin magnetic moment of an electron is determined by:

$$\boldsymbol{\mu} = -2\frac{e}{2m_e}\mathbf{S},$$

The energy of interaction is determined by:

$$E_{mag} = -\boldsymbol{\mu} \cdot \mathbf{B} = \frac{e^2}{4\pi\epsilon_0 m_e^2 c^2 r^3}\mathbf{L} \cdot \mathbf{S}.$$

This calculation has precisely determined the exact size of the spin-orbit interaction

(Das et al., 2023). By including the electron's acceleration, a more precise computation incorporates a factor of one-half, resulting in:

$$E_{mag} = \frac{e^2}{8\pi\epsilon_0 m_e^2 c^2 r^3}\mathbf{L}\cdot\mathbf{S}.$$

(35)

It is important to highlight two fundamental characteristics of the spin-orbit interaction:

a. The spin-orbit interaction is relativistic, as evidenced by its equivalent correction to the relativistic correction of kinetic energy described in Eqn. (33):

$$\langle E_{rel}\rangle \approx -\frac{1}{8}\alpha^4 m_e c^2.$$

By utilizing the fine structure constant, denoted as 'a' in Eqn. (32), one can rephrase Eqn. (35) to express the spin-orbit energy (Kaltsoyannis, 1997):

$$E_{mag} = \alpha\frac{\hbar}{2m_e^2 c}\frac{\mathbf{L}\cdot\mathbf{S}}{r^3}.$$

Rewriting the Bohr radius a_0:

$$a_0 = \frac{1}{\alpha}\frac{\hbar}{m_e c},$$

The spin-orbit energy of a low-lying hydrogen atom state is about equal to:

$$\langle E_{mag}\rangle \approx \alpha^4 m_e c^2.$$

b. The spin-orbit interaction dictates that quantum states of the hydrogen atom with a specific energy possess precise values for J^2, L^2, and S^2. This can be verified by employing (Ermler et al., 1988):

$$\mathbf{J}\cdot\mathbf{J} = (\mathbf{L}+\mathbf{S})\cdot(\mathbf{L}+\mathbf{S}) = L^2 + S^2 + 2\mathbf{L}\cdot\mathbf{S}$$

To show that the spin-orbit interaction is proportional to:

$$\mathbf{L}\cdot\mathbf{S} = \frac{J^2 - L^2 - S^2}{2}.$$

These states will have well-defined energy due to the spin-orbit interaction if they have well-defined orbital (Eickerling et al., 2007), spin, and total angular momenta, as shown by the magnitudes of these quantities:

$$S = \sqrt{s(s+1)}\hbar, \quad L = \sqrt{l(l+1)}\hbar \quad \text{and} \quad J = \sqrt{j(j+1)}\hbar,$$

l, s, and j are the quantum numbers being discussed. Eqn. (6) expresses the equation that governs the combination of spin and orbital angular momenta (Yoo et al., 2009). It states that the variable j may take on several values:

$$j = l + s,\ l + s - 1, \ldots,\ |l - s|.$$

.

In the absence of spin-orbit interaction, all hydrogen atom states with a principal quantum number (n) = 2 would possess identical energy levels:

$$E_2 = -\frac{E_R}{2^2} = -\frac{\alpha^2}{8} m_e c^2,$$

E_R represents the Rydberg energy, while a denotes the fine structure constant (Lamata et al., 2007). The energies are determined by:

$$E(2s_{1/2}) = E_2 - \tfrac{5}{64}\,\alpha^4 m_e c^2,$$

$$E(2p_{1/2}) = E_2 - \tfrac{5}{64}\,\alpha^4 m_e c^2,$$

$$E(2p_{3/2}) = E_2 - \tfrac{1}{64}\,\alpha^4 m_e c^2.$$

We note that the difference in energies of the and states can be verified by observing a small difference in the wavelengths of the radiation emitted by the transitions and . We also note that the and 2 states are predicted to have the same energy, but that this degeneracy is removed by a small effect called the Lamb shift which arises from the quantum field properties of the electromagnetic field."

ACTIVITY 6.1

Objective: To deepen understanding of key concepts related to the hydrogen atom through interactive exploration.

Materials Needed:

- Computer or tablet with internet access
- Pen and paper
- Optional: Access to a simulation or visualization tool for atomic structures

Instructions:

a. Research and Reflect:
   - Use online resources to research the classical and quantum mechanical concepts discussed in the chapter.
   - Reflect on how central potentials influence particle motion and the implications for atomic structure.

b. Wave Function Exploration:
   - Explore the components of the hydrogen atom's wave function. Consider how different quantum numbers affect the wave function and consequently the energy levels of the atom.
   - Write down your observations and insights regarding the wave function's role in describing the hydrogen atom.

SUMMARY

- Beginning with classical mechanics and progressing up to quantum mechanics, the chapter titled "The Hydrogen Atom" delves into the fundamental concepts of central potentials in the field of physics. In the first step of the analysis, the behavior of particles under central potentials in classical mechanics is investigated, with a specific emphasis placed on the orbits and trajectories of the particles. The wave function is a notion that is introduced in quantum mechanics to characterize particles in central potentials. This concept makes it possible to conduct an in-depth investigation of the hydrogen atom.
- Within the context of the quantum framework, this chapter dives into the energy levels and eigenfunctions of the hydrogen atom, both of which are fundamental for understanding the structure of atoms. This research provides a comprehensive account of the spatial distribution of electron probability densities. It explores how quantum states influence the size and arrangement of atoms. This article examines radiative transitions between discrete energy levels and offers valuable insights into the mechanisms via which atoms emit and absorb photons. The chapter also investigates the relevance of relativistic effects at high velocities or in heavy nuclei, as well as the impact of the decreased mass effect on the behavior of particles in central potentials. In addition, the chapter investigates the impact of the reduced mass effect. These components contribute to a comprehensive understanding of atomic physics and offer insights into wider applications in quantum mechanics and atomic spectroscopy (both of which are related to atomic physics).

REVIEW QUESTIONS

1. How does a central potential influence particle motion in classical and quantum mechanics?
2. What are the key components of the hydrogen atom's wave function, and how do they relate to its energy levels?
3. What role does the principal quantum number n play in determining the size, energy, and spatial distribution of electron orbitals in hydrogen?
4. How do radiative transitions occur in the hydrogen atom, and what is their connection to photon emission or absorption and changes in electron energy levels?
5. Explain how the concept of reduced mass affects the dynamics of the electron relative to the proton in a hydrogen atom.
6. Briefly discuss the adjustments to energy levels in the hydrogen atom due to relativistic effects, and identify the conditions under which these effects become significant.

MULTIPLE CHOICE QUESTIONS

1. **What does the quantum mechanics of a particle in a central potential describe?**
   a. Motion of particles under a force directed towards a fixed point
   b. Motion of particles in a magnetic field
   c. Motion of particles under varying gravitational fields
   d. Motion of particles in a random environment
2. **What is the principal quantum number in the context of the hydrogen atom?**
   a. Number of neutrons in the nucleus
   b. Number of protons in the nucleus
   c. Number that determines the size and energy of an orbital
   d. Number of electrons in the atom
3. **Which phenomenon describes the emission or absorption of light by atoms as electrons change energy levels?**
   a. Quantum tunneling
   b. Photoelectric effect
   c. Radiative transitions
   d. Compton scattering
4. **What does the reduced mass effect refer to in the context of the hydrogen atom?**
   a. The difference in mass between the electron and the proton
   b. The effective mass of the electron in the hydrogen atom
   c. The change in mass due to relativistic effects
   d. The combined mass of the electron and the proton
5. **How do relativistic effects impact the energy levels of the hydrogen atom?**
   a. They cause the energy levels to converge
   b. They cause the energy levels to split
   c. They have no effect on the energy levels
   d. They shift the energy levels slightly
6. **What is the shape of the orbitals in a hydrogen atom described by?**
   a. Principal quantum number
   b. Orbital angular momentum quantum number
   c. Magnetic quantum number
   d. Spin quantum number

Answers to Multiple Questions

1. (a); 2. (c); 3. (c); 4. (b); 5. (d); 6. (b)

REFERENCES

1. Aad, G., Abbott, B., Abdallah, J., Abdelalim, A. A., Abdesselam, A., Abdinov, O., & Barashkou, A. (2012). Observation of a new χ_b state in radiative transitions to ϒ(1S) and ϒ(2S) at ATLAS. Physical Review Letters, 108(15), 152001.
2. Aashamar, K., Luke, T. M., & Talman, J. D. (1978). Optimized central potentials for atomic ground-state wavefunctions. Atomic Data and Nuclear Data Tables, 22(5), 443–472.
3. Bander, M., & Itzykson, C. (1966). Group theory and the hydrogen atom (I). Reviews of Modern Physics, 38(2), 330.
4. Barton, G. (1986). Quantum mechanics of the inverted oscillator potential. Annals of Physics, 166(2), 322–363.
5. Bromberg, J. (1976). The concept of particle creation before and after quantum mechanics. Historical Studies in the Physical Sciences, 7, 161–191.
6. Campos, P., Martins, M. G. R., Fernandes, M. C. B., & Vianna, J. D. M. (2018). Quantum mechanics on phase space: The hydrogen atom and its Wigner functions. Annals of Physics, 390, 60–70.
7. Casati, G., Chirikov, B. V., Shepelyansky, D. L., & Guarneri, I. (1987). Relevance of classical chaos in quantum mechanics: The hydrogen atom in a monochromatic field. Physics Reports, 154(2), 77–123.
8. Çetin, A. (2017). Exact energy levels and eigenfunctions of an electron on a nanosphere under the influence of a radial magnetic field. Physica B: Condensed Matter, 523, 92–95.
9. Ching, W. Y., & Huber, D. L. (1982). Numerical studies of energy levels and eigenfunction localization in dilute three-dimensional systems with exponential interactions. Physical Review B, 25(2), 1096.
10. Cisneros, A., & McIntosh, H. V. (1969). Symmetry of the two-dimensional hydrogen atom. Journal of Mathematical Physics, 10(2), 277–286.
11. Colangelo, P., De Fazio, F., & Ozpineci, A. (2005). Radiative transitions of D_sJ*(2317) and D_sJ(2460). Physical Review D—Particles, Fields, Gravitation, and Cosmology, 72(7), 074004.
12. Collas, P. (1981). Equivalent potentials in classical mechanics. Journal of Mathematical Physics, 22(11), 2512–2517.
13. Condon, E. U., & Morse, P. M. (1931). Quantum mechanics of collision processes I. Scattering of particles in a definite force field. Reviews of Modern Physics, 3(1), 43.
14. Das, A., Das, U., & Das, A. K. (2023). Relativistic effects on the chemical bonding properties of the heavier elements and their compounds. Coordination Chemistry Reviews, 479, 215000.
15. Dimopoulos, S., Graham, P. W., Hogan, J. M., & Kasevich, M. A. (2008). General relativistic effects in atom interferometry. Physical Review D—Particles, Fields, Gravitation, and Cosmology, 78(4), 042003.

16. Dirac, P. A. M. (1926). Quantum mechanics and a preliminary investigation of the hydrogen atom. Proceedings of the Royal Society of London. Series A, Containing Papers of a Mathematical and Physical Character, 110(755), 561–579.

17. Donelan, J. M., & Kram, R. (1997). The effect of reduced gravity on the kinematics of human walking: A test of the dynamic similarity hypothesis for locomotion. Journal of Experimental Biology, 200(24), 3193–3201.

18. Dyks, J., Harding, A. K., & Rudak, B. (2004). Relativistic effects and polarization in three high-energy pulsar models. The Astrophysical Journal, 606(2), 1125.

19. Eickerling, G., Mastalerz, R., Herz, V., Scherer, W., Himmel, H. J., & Reiher, M. (2007). Relativistic effects on the topology of the electron density. Journal of Chemical Theory and Computation, 3(6), 2182–2197.

20. Ermler, W. C., Ross, R. B., & Christiansen, P. A. (1988). Spin-orbit coupling and other relativistic effects in atoms and molecules. In Advances in Quantum Chemistry (Vol. 19, pp. 139–182). Academic Press.

21. Finkelstein-Shapiro, D., Mante, P. A., Sarisozen, S., Wittenbecher, L., Minda, I., Balci, S., & Zigmantas, D. (2021). Understanding radiative transitions and relaxation pathways in plexcitons. Chem, 7(4), 1092–1107.

22. Frosini, P. (1992). Measuring shapes by size functions. In Intelligent Robots and Computer Vision X: Algorithms and Techniques (Vol. 1607, pp. 122–133). SPIE.

23. Gnanasangeetha, G. (2022). Eigenfunction method for splitting of energy levels in quantum mechanics. International Journal of Modern Physics B, 36(23), 2250150.

24. Gordeyev, A. N., & Chhajlany, S. C. (1997). One-dimensional hydrogen atom: A singular potential in quantum mechanics. Journal of Physics A: Mathematical and General, 30(19), 6893.

25. Grant, I. P. (1974). Gauge invariance and relativistic radiative transitions. Journal of Physics B: Atomic and Molecular Physics, 7(12), 1458.

26. Gu, P. G., Lin, D. N., & Bodenheimer, P. H. (2003). The effect of tidal inflation instability on the mass and dynamical evolution of extrasolar planets with ultrashort periods. The Astrophysical Journal, 588(1), 509.

27. Hellings, R. W. (1986). Relativistic effects in astronomical timing measurements. Astronomical Journal, 91(March 1986), 650–659.

28. Hendi, A., Henn, J., & Leonhardt, U. (2006). Ambiguities in the scattering tomography for central potentials. Physical Review Letters, 97(7), 073902.

29. Holas, A., & March, N. H. (1990). A generalization of the Runge-Lenz constant of classical motion in a central potential. Journal of Physics A: Mathematical and General, 23(5), 735.

30. Huang, K., & Rhys, A. (1950). Theory of light absorption and non-radiative transitions in F-centres. Proceedings of the Royal Society of London. Series A. Mathematical and Physical Sciences, 204(1078), 406–423.

31. Huang, S. W., & Carrington, T. (2000). A new iterative method for calculating energy levels and wave functions. The Journal of Chemical Physics, 112(20), 8765–8771.

32. Jelic, V., & Marsiglio, F. (2012). The double-well potential in quantum mechanics: A simple, numerically exact formulation. European Journal of Physics, 33(6), 1651.

33. Kalendová, A. (2003). Effects of particle sizes and shapes of zinc metal on the properties of anticorrosive coatings. Progress in Organic Coatings, 46(4), 324–332.

34. Kaltsoyannis, N. (1997). Relativistic effects in inorganic and organometallic chemistry. Journal of the Chemical Society, Dalton Transactions, (1), 1–12.

35. Kasiske, B. L., Ma, J. Z., Louis, T. A., & Swan, S. K. (1995). Long-term effects of reduced renal mass in humans. Kidney International, 48(3), 814–819.

36. Kram, R., Domingo, A., & Ferriss, D. P. (1997). Effect of reduced gravity on the preferred walk–run transition speed. Journal of Experimental Biology, 200(4), 821–826.

37. Krupke, W. (1971). Radiative transition probabilities within the 4f 3 ground configuration of Nd: YAG. IEEE Journal of Quantum Electronics, 7(4), 153–159.

38. Kryukov, A. (2013). Functional methods underlying classical mechanics, relativity, and quantum theory. In Journal of Physics: Conference Series, 437(1), 012020.

39. Kubo, R., & Toyozawa, Y. (1955). Application of the method of generating function to radiative and non-radiative transitions of a trapped electron in a crystal. Progress of Theoretical Physics, 13(2), 160–182.

40. Lamata, L., León, J., Schätz, T., & Solano, E. (2007). Dirac equation and quantum relativistic effects in a single trapped ion. Physical Review Letters, 98(25), 253005.

41. Lamb Jr, W. E., & Retherford, R. C. (1951). Fine structure of the hydrogen atom. Part II. Physical Review, 81(2), 222.

42. Lass, E. A., Sauza, D. J., Dunand, D. C., & Seidman, D. N. (2018). Multicomponent γ'-strengthened Co-based superalloys with increased solvus temperatures and reduced mass densities. Acta Materialia, 147, 284–295.

43. MacMahon, S. W., Wilcken, D. E., & Macdonald, G. J. (1986). The effect of weight reduction on left ventricular mass. New England Journal of Medicine, 314(6), 334–339.

44. Marcus, Y. (1994). A simple empirical model describing the thermodynamics of hydration of ions of widely varying charges, sizes, and shapes. Biophysical Chemistry, 51(2–3), 111–127.

45. Mielnik, B. (1974). Generalized quantum mechanics. Communications in Mathematical Physics, 37, 221–256.

46. Mills, R. L. (2000). The hydrogen atom revisited. International Journal of Hydrogen Energy, 25(12), 1171–1183.

47. Mills, R. L. (2005). The fallacy of Feynman's and related arguments on the stability of the hydrogen atom according to quantum mechanics. In Annales de la Foundation Louis de Broglie (Vol. 30, No. 2, pp. 129–149). Foundation Louis de Broglie.

48. Mirlin, A. D. (2000). Statistics of energy levels and eigenfunctions in disordered systems. Physics Reports, 326(5–6), 259–382.

49. Muralidhar, K. (2021). Common physical theory for quantum/classical particles in complex vector space. Quantum Studies: Mathematics and Foundations, 8(2), 249–259.

50. Neuhauser, D. (1990). Bound state eigenfunctions from wave packets: Time→ energy resolution. The Journal of Chemical Physics, 93(4), 2611–2616.
51. Nishiyama, Y. (1965). On a parallelism between classical mechanics and quantum mechanics. Progress of Theoretical Physics, 34(2), 299–311.
52. Paradoksov, P. (1966). How quantum mechanics helps us understand classical mechanics. Uspekhi Fizicheskikh Nauk, 89(4), 707–709.
53. Parfitt, D. G. W., & Portnoi, M. E. (2002). The two-dimensional hydrogen atom revisited. Journal of Mathematical Physics, 43(10), 4681–4691.
54. Pauling, L. (1928). The application of the quantum mechanics to the structure of the hydrogen molecule and hydrogen molecule-ion and to related problems. Chemical Reviews, 5(2), 173–213.
55. Pyykkö, P. (2012). Relativistic effects in chemistry: More common than you thought. Annual Review of Physical Chemistry, 63(1), 45–64.
56. Rizov, V. A., Sazdjian, H., & Todorov, I. T. (1985). On the relativistic quantum mechanics of two interacting spinless particles. Annals of Physics, 165(1), 59–97.
57. Salzemann, C., Lisiecki, I., Brioude, A., Urban, J., & Pileni, M. P. (2004). Collections of copper nanocrystals characterized by different sizes and shapes: Optical response of these nanoobjects. The Journal of Physical Chemistry B, 108(35), 13242–13248.
58. Schwerdtfeger, P. (2002). Relativistic effects in properties of gold. Heteroatom Chemistry: An International Journal of Main Group Elements, 13(6), 578–584.
59. Simon, H. A. (1954). The axiomatization of classical mechanics. Philosophy of Science, 21(4), 340–343.
60. Stevenson, P. M. (1984). Gaussian effective potential: Quantum mechanics. Physical Review D, 30(8), 1712.
61. Stewart, R. F., Davidson, E. R., & Simpson, W. T. (1965). Coherent X-ray scattering for the hydrogen atom in the hydrogen molecule. The Journal of Chemical Physics, 42(9), 3175–3187.
62. Trushechkin, A. S., & Volovich, I. V. (2009). Functional classical mechanics and rational numbers. p-Adic Numbers, Ultra metric Analysis, and Applications, 1, 361–367.
63. Van Lenthe, E., Ehlers, A., & Baerends, E. J. (1999). Geometry optimizations in the zero-order regular approximation for relativistic effects. The Journal of Chemical Physics, 110(18), 8943–8953.
64. Voth, G. A. (1996). Path-integral centroid methods in quantum statistical mechanics and dynamics. Advances in Chemical Physics, 135–218.
65. Yamamoto, A., Honma, R., Sumita, M., & Hanawa, T. (2004). Cytotoxicity evaluation of ceramic particles of different sizes and shapes. Journal of Biomedical Materials Research Part A: An Official Journal of The Society for Biomaterials, The Japanese Society for Biomaterials, and The Australian Society for Biomaterials and the Korean Society for Biomaterials, 68(2), 244–256.
66. Yoo, J., Fitzpatrick, A. L., & Zaldarriaga, M. (2009). New perspective on galaxy clustering as a cosmological probe: General relativistic effects. Physical Review D—Particles, Fields, Gravitation, and Cosmology, 80(8), 083514.

CHAPTER 7

Scattering Theory

LEARNING OBJECTIVES

After studying this chapter, you will be able to:

1. Understand the fundamental concepts of scattering in quantum mechanics and its importance.
2. Define and calculate the differential and total cross-sections for various scattering processes.
3. Distinguish between laboratory and center-of-mass frames of reference and perform relevant transformations.
4. Relate the quantum state function to scattering phenomena and solve the Schrödinger equation for scattering problems.
5. Analyze scattering by a spherical potential and understand the role of phase shifts.
6. Apply the partial wave analysis to determine the cross sections for different types of scattering potentials..

KEY TERMS FROM THIS CHAPTER

center-of-mass frame
elastic collision
laboratory frame
phase shifts
scattering potential
differential cross section
inelastic collision
partial wave analysis
quantum state function
total cross section

UNIT INTRODUCTION

In quantum mechanics, scattering does not provide deterministic predictions for individual measurements. Instead, it offers statistical distributions or probability for all potential possibilities. Scattering plays a crucial role, surpassing the significance portrayed in the above figure (Rauch, & Taylor, 1975). A substantial amount of knowledge regarding particle interactions is obtained through scattering experiments. Scattering theory is a complex subject that has been extensively covered in numerous works. However, this chapter will merely provide an overview of the fundamental concepts (Lax & Phillips, 1971).

This chapter is dedicated to the study of scattering theory in quantum mechanics, a fundamental concept for comprehending the interactions between particles in subjects such as particle physics, nuclear physics, and the physics of condensed matter (Strauss, 1981). The chapter discusses the mathematical representation and understanding of scattering processes. Important ideas encompass total cross sections and differential, a lab, and center-of-mass reference frames, followed by partial wave analysis. Through a thorough examination of these domains, readers will acquire a strong fundamental understanding of scattering theory including its practical implementations (Lundstrom, 1997).

7.1. CROSS SECTION

The scattering pattern of particles in a specific process is characterized by a differential cross-section, which describes the angular distribution of the dispersed particles. Assume that a flow of J_i particles for every unit area for each unit of time is coming towards the target. The number of particles scattered into a narrow cone of solid angle $d\Omega$, centered on the direction defined by polar angles θ and φ with respect to the incident flux, is directly proportional to the incident flux J_i and the angular opening $d\Omega$ of the cone. Therefore, it can be expressed as . The factor of proportionality is referred to as the differential cross section (Kesmodel, 2018).

Assume that there is a particle detector positioned at coordinates (θ, φ), which is situated at a significant distance r from the target in order to be outside the path of the incoming beam (Koch & Motz, 1959) (Figure 7.1).

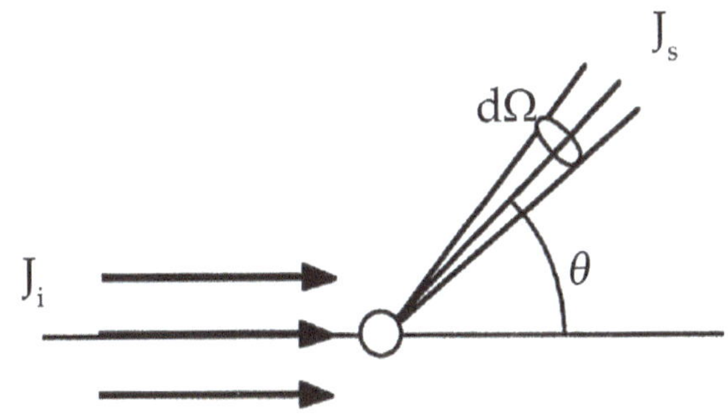

Figure 7.1. Illustration of the differential cross section.

Source: Leslie E, Creative Commons License.

If it delimits the solid angle $d\Omega$, the equation tells that it will get scattered particles in a unit time. The ratio of this number to the detector area, the flux at the detectoris obtained for the scattered particles.

Thus,

$$\sigma(\theta, \phi) = \frac{r^2 J_s}{J_i}, \quad (1)$$

It is evident that the object has the measurements of an area. The value of J_s is not affected by the distance r between the target and the detector, as J_s varies inversely with the square of r. This equation is advantageous since the fluxes J_s and J_i are quantifiable variables that may be measured and computed theoretically (Carlson et al., 2009).

The overall cross-section is obtained by integrating over all scattering directions,

$$\sigma = \int \sigma(\theta, \phi)\, d\Omega$$

$$= \int_0^{2\pi} \int_0^{\pi} \sigma(\theta, \phi) \sin\theta \, d\theta \, d\phi \,. \qquad (2)$$

7.1.1. Laboratory and Center-of-Mass Frames

When defining σ(θ,φ) before, an assumption that the goal was stationary was made. This statement is not always accurate, as the combined momentum of the bullet and targeted particles remains constant. In order to conduct a theoretical analysis, it is most advantageous to utilize a frame of reference where the center of mass of both particles remains stationary. The characterization of the scattering event is subsequently symmetrical between the target and the projectile. In the above formulas, the terms "distance r" and "direction (θ,φ)" represent the relative distancing of the projectile and the target. Additionally, (θ, φ) represents the trajectory of the scattered projectile originating from the stationary center of mass, whereas the rebound of the target particle occurs in the reverse direction (π − θ, φ + π). The calculation of scattering cross-sections is often performed in the center-of-mass (CM) frame (Martinez et al., 1999).

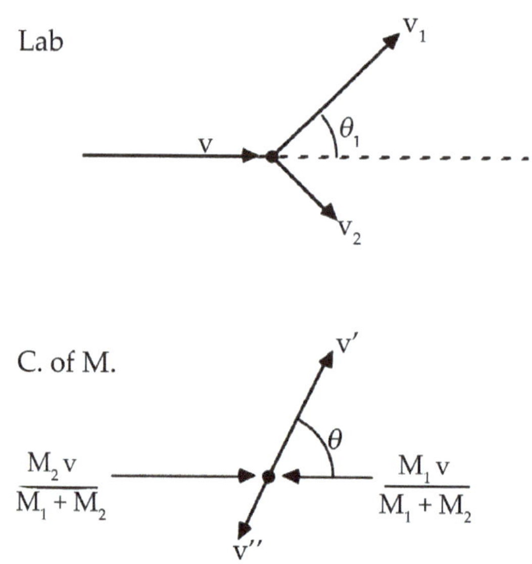

Figure 7.2. Illustration of Scattering event in the laboratory frame (top), and in the center-of-mass frame (bottom).

Source: Belanine, Creative Commons License.

A lab frame of reference is used to generate experimental data, where the targeted particle is initially stationary (see Figure 7.2). In the laboratory frame, there is an initial configuration where there is a projectile particle having mass M_1 as well as velocity v_1, and another targeted particle having mass M_2 that is stationary. The CM velocity relative to the laboratory frame is expressed by (Setterfield et al., 2018). To convert from the coordinates of the laboratory to the reference frame where the center of mass is stationary, it is needed to deduct V_0 from all velocities. Thus, the initial velocity of the projectile is and the initial target velocity is Figure 7.2 makes it abundantly clear that the ultimate velocity and trajectory of the projectile in both the reference frames are related by a relationship that can be expressed as:

$$v_1 \cos\theta_1 = v' \cos\theta + V_0,$$
$$v_1 \sin\theta_1 = v' \sin\theta. \quad (3)$$

The ratios of these two equations give;

$$\tan\theta_1 = \frac{\sin\theta}{\cos\theta + \beta}, \quad \text{with } \beta = \frac{V_0}{v'}. \quad (4)$$

There is no change in the velocity of the particles in relation to the CM when they collide with an elastic material, and so

Thus;

In a typical inelastic collision, the internal energy associated with the particles can vary, meaning that their total kinetic energy is not necessarily preserved. During a rearrangement collision involving composite particles, there is an exchange of mass among the particles. (Instances include nuclear processes, chemical reactions involving molecules, and charge exchange involving atoms and ions (Carpentier et al., 2016). Assume that the masses of the particles coming in are denoted as M_1 and M_2, while the masses of the particles going out are denoted as M_3 and M_4. Considering non-relativistic kinematics, . Demonstrably:

$$\beta = \left\{ \frac{M_1\, M_3\, E}{M_2\, M_4\, (E + Q)} \right\}^{1/2}. \quad (5)$$

In this context, M_1 represents the projectile mass, M_2 represents the starting point mass of the target, M_3 represents the mass of the observed particle (whose direction is denoted as θ in the center of mass frame), and M_4 represents the mass of the recoil particle, which is typically not detected. The starting kinetic energy in the center of mass frame is while Q represents the quantity of internal energy transformed into kinetic energy during the reaction. The upper limit of an elastic collision is determined by setting Q = 0 and . The relationship between the differential cross sections in the lab and center of mass (CM) frames can be established by considering that the quantity of particles distributed into a specific cone must be equal in both coordinate systems (Wang & Marshall, 1990).

The incidence flux, which refers to the number of particles per unit area per unit time, is constant in both frames:

$$\sigma_1(\theta_1, \phi_1) \sin\theta_1 \, d\theta_1 \, d\phi_1 = \sigma(\theta, \phi) \sin\theta \, d\theta \, d\phi \, . \tag{6}$$

The relationship between θ_1 and θ is defined by Eqn. (4), therefore it is evident that $\varphi_1 = \varphi$. By applying algebraic operations, it can be deduced that the cross-section in the lab frame is expressed as:

$$\sigma_1(\theta_1, \phi_1) = \sigma(\theta, \phi) \frac{(1 + \beta^2 + 2\beta \cos\theta)^{3/2}}{|1 + \beta \cos\theta|} \tag{7}$$

The cross-section must be consistent between the two frames of reference, as the total number of scattered particles remains unaffected by the choice of coordinate system (Alba et al., 2002).

7.1.2. The Quantum State Function in Scattering

Quantum mechanics was not mentioned specifically in the preceding section. These discussions focused on the formal explanations of cross-sections in relation to the number of scattered particles, as well as the change of velocity vectors across various frames of reference. Those results are unaffected by the disparities between traditional and quantum physics. It is now needed to establish the connection between those definitions and the quantum state function (Enss, 1983).

The Hamiltonian governing the dynamics of a system consisting of two particles that interact with each other is given by:

$$H = -\frac{\hbar^2}{2M_1} \nabla_1{}^2 - \frac{\hbar^2}{2M_2} \nabla_2{}^2 + V(\mathbf{x}_1 - \mathbf{x}_2) \, , \tag{8}$$

The first term includes derivatives with regard to $\mathbf{x}_1$, while the second term includes derivatives at $\mathbf{x}_2$. (To simplify, the internal degrees of freedom (DOF) of the two particles have not been shown). The initial procedure involves transforming the parameters from the coordinates of the specific particles to the center of mass (CM) and relative coordinates,

$$\mathbf{X} = \frac{M_1 \mathbf{x}_1 + M_2 \mathbf{x}_2}{M_1 + M_2} \, , \tag{9}$$

$$\mathbf{r} = \mathbf{x}_1 - \mathbf{x}_2 \, .$$

This transformation was achieved using a canonical transformation of the momentum and position operators. However, it can be accomplished more easily by inserting the new variables Eqn. (9) in the differential operators of Eqn. (8) (Blum, 2017). The outcome is:

$$H = -\frac{\hbar^2}{2(M_1 + M_2)} \nabla x^2 - \frac{\hbar^2}{2\mu} \nabla^2 + V(r) . \quad (10)$$

The first term refers to the kinetic energy of the center of mass, which is currently not relevant. The second term represents the kinetic energy associated with the relative motion. Its derivatives are calculated with respect to the relative coordinate, r. The reduced mass is given as, The second factor fulfills:

$$-\frac{\hbar^2}{2\mu} \nabla^2 \psi(\mathbf{r}) + V(r) \psi(\mathbf{r}) = E \psi(\mathbf{r}) , \quad (11)$$

E is the energy linked to the relative velocity of the two particles in the center-of-mass frame.

The suitable boundary condition for Eqn. (11) is established based on the experimental conditions, as seen in Figure 7.1, which are employed to define the differential cross-section. There is a requirement for an incident flux J_i that moves from the point of origin to the target, as well as a scattered flux J_s that spreads out in all directions. The Eqn. (11) does not account for the particle source (Kriman et al., 1987). Therefore, the boundary condition must be implemented to determine the value of the incident flux. Hence, it is necessary for the solution of Eqn. (11) to take the specific form:

$$\psi(\mathbf{r}) = \psi_i(\mathbf{r}) + \psi_s(\mathbf{r}) , \quad (12)$$

whereas the "incident wave" symbolizes the incident beam flux, and is an outward "scattered wave."

The quantum state does not provide information about the precise location of the incoming particle, but instead, it provides the probability density, , for the particle to be at a distance r from the target. Similarly, the state does not depict the precise flow of particles, but rather the probability flow, which represents the overall likelihood per unit time that a particle passes through a unit region (Tannor & Weeks, 1993). When considering this specific case, one can express the probability flux as shown in Eqn. (11):

$$\mathbf{J} = \frac{\hbar}{\mu} \mathrm{Im}(\psi^* \boldsymbol{\nabla} \psi) . \quad (13)$$

The incident beam is given by , which has a uniform flux, . If is finite, then for higher values of r, Eqn. (11) will be reduced to the free particle equation, and one can anticipate ψ(r) to be asymptotically equal to a solution of the free particle equation. Therefore, a solution satisfying the asymptotic boundary condition of Eqn. (11) is needed:

$$\psi(\mathbf{r}) \sim A\left\{e^{i\mathbf{k}\cdot\mathbf{r}} + \frac{f(\theta,\phi)e^{ikr}}{r}\right\}$$

(14)

By replacing the flow rates of these two terms into Eqn. (1) (Zhang et al., 1992), the differential cross section is obtained:

$$\sigma(\theta,\phi) = |f(\theta,\phi)|^2 .$$

(15)

Hence, the resolution of a scattering problem is simplified to identifying the asymptotic characteristics of the scattering state function. A (amplitude) is inconsequential and is typically assigned a value of 1 for the sake of convenience. By disregarding the intrinsic DOF of the particles, the solution has been made implicitly limited to the scenario of elastic scattering. The outcome Eqn. (15) will be altered when considering inelastic scattering (Hunziker & Sigal, 2000).

The flux is not computed by replacing Eqn. (14) with Eqn. (13). Instead, distinct fluxes were computed for the two components of Eqn. (14), inadvertently excluding crucial cross terms. The statement is not an error, but rather an acknowledgment that the incident beam has to be a certain width, as shown in Figure 7.1. Thus, , within the incident beam, and elsewhere. At the detector, if it is outside of the incident beam. In practice, it is always necessary to include the cross terms involving both ψ_i (r) and ψ_s (r) in J. This is done for obvious reasons. However, if this were not the case, these cross-terms would need to be included.

The majority of scattering theory focuses on finding the solution to Eqn. (11) while taking into account a boundary condition such as Eqns. (12) or (14). When focusing on the specific aspects of that solution, it is natural to consider ψ(r) as the description of the movement of the scattered particle, while seeing the target particle as a stationary force center that sustains the potential V(r). Technically speaking, the interpretation is incorrect. Eqn. (11) provides a comprehensive account of the relative movement between the two particles. This description is entirely symmetrical in relation to both particles, except for the reversal of the sign of r when they are swapped. The state function ψ(r) characterizes the quantum-mechanical attributes of both the target particle and the incident particle (Mekhov & Ritsch, 2009).

7.2. SCATTERING BY A SPHERICAL POTENTIAL

This section provides a comprehensive analysis of particle scattering, specifically focusing on particles that interact through a potential V(r) that is only dependent on the distance between the particles. As this contact does not cause any changes in the particles' internal degrees of freedom, one can disregard them. Exclusively elastic scattering is feasible within this model (Zhang & Ho, 2018).

For the sake of convenience, Eqn. (11), which describes the condition of relative motion between the two particles, shall be restated:

$$\nabla^2\psi(\mathbf{r}) + \left[k^2 - U(r)\right]\psi(\mathbf{r}) = 0\,, \qquad (16)$$

Where; and

$$\psi(\mathbf{r}) = \sum_{\ell m} a_{\ell m}\, Y_\ell{}^m(\theta,\phi)\, \frac{u_\ell(r)}{r}\,, \qquad (17)$$

Where the radial functions fulfill the equation:

$$\frac{d^2 u_\ell(r)}{dr^2} + \left[k^2 - U(r) - \frac{\ell(\ell+1)}{r^2}\right] u_\ell(r) = 0\,. \qquad (18)$$

It is necessary to ascertain the asymptotic trajectory of the radial function $u_l(r)$ at this point. For values of r that are sufficiently big, the U and l components can be considered insignificant as compared to k^2. This implies that the asymptotic shape of the radial function must be . In order to validate this intuitive yet nonrigorous estimation, an experiment was conducted:

$$u_\ell(r) = e^{h(r)}\, e^{\pm ikr}\,, \qquad (19)$$

At large values of r, anticipate that the first exponential will exhibit a gradual change, in contrast to the second exponential (Bobbert & Vlieger, 1986). By substituting Eqn. (19) into Eqn. (18), obtain a differential equation for the function h(r),

$$h''(r) + \left[h'(r)\right]^2 \pm 2ikh'(r) = U(r) + \frac{\ell(\ell+1)}{r^2}\,, \qquad (20)$$

Where the primes denote differentiation. The above equation is a first-order differential equation for h'(r) as it does not involve h(r). If the function U(r) decreases fast as r approaches infinity, following at least a rate of r^2, then the third component on the left side of the equation will have the most influence. In this case, the derivative of h(r), denoted as h'(r), will approach zero at a rate of r^2. At large values of r, the function h(r) can be expressed as , and as r approaches infinity, h(r) becomes a constant. The initial estimate is accurate in this scenario, and $u_l(r)$ does follow the exponential function e±ikr for large values of r, which aligns with the asymptotic form Eqn. (14) (Charalambopoulos & Dassios, 1999).

This argument also demonstrates why the Coulomb potential, which is necessitates unique handling. Given this scenario, it is evident from Eqn. (20) that the derivative of h with respect to r, denoted as h'(r), decreases proportionally to r^{-1} as r becomes larger. Consequently, h(r) increases logarithmically with r and does not approach a finite value as r approaches infinity. As a result, the initial exponential term in Eqn. (19) does not approach a constant value as r becomes larger. Therefore, it is not possible to achieve the asymptotic form Eqn. (14). Citations regarding the specific requirements for the Coulomb potential can be found at the conclusion of this chapter. From now on, the discussion will focus exclusively on short-range potentials (Videen, 1991). This refers to potentials that decrease at huge distances, namely at a rate of at least r^{-2}.

7.2.1. Phase Shifts

It is advantageous to examine a narrower range of potentials that either disappear or become insignificant in comparison to the term when r is greater than a. Subsequently, it will be demonstrated how the main findings can be extended to encompass any potential with a limited range (Hughes et al., 2007).

Let's address the issue of solving Eqn. (16) while considering the boundary condition Eqn. (14). Assuming the scattering potential is completely absent, the only solution (without normalization) would be:

$$\psi_i = e^{i\mathbf{k}\cdot\mathbf{r}} = \sum_{\ell}(2\ell+1)\, i^{\ell}\, j_{\ell}(kr)\, P_{\ell}(\cos\theta)\,, \tag{21}$$

The function j_l represents a spherical Bessel function, P_l represents a Legendre polynomial, and θ represents the angle between k and r. Let's express the solution of Eqn. (16) with the scattering potential in the following format:

$$\psi = \sum_{\ell}(2\ell+1)\, i^{\ell}\, A_{\ell}\, R_{\ell}(r)\, P_{\ell}(\cos\theta) \tag{22}$$

where fulfills the partial wave equation,

$$\frac{1}{r^2}\frac{d}{dr}r^2\frac{d}{dr}R_{\ell}(r) + \left[k^2 - U(r) - \frac{\ell(\ell+1)}{r^2}\right]R_{\ell}(r) = 0\,. \tag{23}$$

In the absence of the potential U(r), the equation met by the spherical Bessel functions would be applicable. Refer to Morse and Feshbach for a comprehensive list of the numerous formulas and identities that these functions satisfy (Hibino et al., 1997). Therefore, when r is greater than a and U(r) equals 0, the solution of Eqn. (23) must be expressed as a linear combination of these two Bessel functions, denoted as:

$$R_\ell(r) = \cos(\delta_\ell)\, j_\ell(kr) - \sin(\delta_\ell)\, n_\ell(kr)\,, \quad (r \geq a)\,. \tag{24}$$

The asymptotic forms for are:

$$j_\ell(kr) \to \frac{\sin(kr - \frac{1}{2}\pi\ell)}{kr}\,, \tag{25}$$

$$n_\ell(kr) \to \frac{-\cos(kr - \frac{1}{2}\pi\ell)}{kr}\,,$$

The corresponding limit is:

$$R_\ell(r) \to \frac{\sin(kr - \frac{1}{2}\pi\ell + \delta_\ell)}{kr}\,. \tag{26}$$

Assuming the scattering potential was precisely zero, the validity of the form Eqn. (24) would extend to the point where r equals zero. However, the function $n_1(kr)$ exhibits a singularity of r_1 at r = 0, which is not permissible for a state function. Therefore, δA must be equal to 0 if U (r) = 0 for all values of r. By evaluating the asymptotic limit for the zero scattering solution, j_1 (kr), with Eqn. (26), it becomes evident that the sole impact of the short-range scattering potential at large r is the alteration of the phase shift of the radial function by δA. Given that the differential cross-section is solely dictated by the asymptotic shape of the state function, it can be concluded that the cross-section can be expressed in terms of these phase shifts (Fushman et al., 2008).

By substituting the series Eqns. (22) and (21) into Eqn. (14) with A = 1, and replacing the Bessel functions with their asymptotic limits, one can get the following result:

$$\sum_\ell (2\ell + 1)\, i^\ell\, P_\ell(\cos\theta)\, A_\ell\, \frac{\sin(kr - \frac{1}{2}\pi\ell + \delta_\ell)}{kr} \tag{27}$$

Remember

Phase shifts are a critical component in understanding scattering. They result from the interaction of the incident wave with the potential and affect the interference pattern of the scattered waves.

$$= \sum_{\ell} (2\ell + 1)\, i^{\ell}\, P_{\ell}(\cos\theta)\, \frac{\sin(kr - \frac{1}{2}\pi\ell)}{kr} + f(\theta, \phi)\, \frac{e^{ikr}}{r}\,.$$

Implying equating the coefficients of :

$$\sum_{\ell} (2\ell + 1)\, i^{\ell}\, P_{\ell}(\cos\theta)\, A_{\ell}\, \exp\left(i\tfrac{1}{2}\pi\ell - i\delta_{\ell}\right)$$

$$= \sum_{\ell} (2\ell + 1) i^{\ell}\, P_{\ell}(\cos\theta)\, \exp\left(i\tfrac{1}{2}\pi\ell\right)\,.$$

Or;

$$A_{\ell} = \exp(i\delta_{\ell})\,. \tag{28}$$

Comparing the coefficients of in Eqn. (27) and using Eqn. (28):

$$f(\theta, \phi) = f(\theta)$$

$$= (2ik)^{-1} \sum_{\ell} (2\ell + 1) i^{\ell}\, \exp\left(-\tfrac{1}{2}\, i\pi\ell\right) [\exp(2i\delta_{\ell}) - 1]\, P_{\ell}(\cos\theta)$$

$$= (2ik)^{-1} \sum_{\ell} (2\ell + 1) [\exp(2i\delta_{\ell}) - 1]\, P_{\ell}(\cos\theta) \tag{29}$$

$$= k^{-1} \sum_{\ell} (2\ell + 1)\, \sin(\delta_{\ell})\, \exp(i\delta_{\ell})\, P_{\ell}(\cos\theta)\,.$$

It is important to observe that this expression remains the same even after substituting . As a result, all scattering effects exhibit periodicity in δ_1 with a period of π, which is different from the expected period of 2π. The current expression for the differential cross section is Eqn. (15):

$$\sigma(\theta, \phi) = \sigma(\theta) = |f(\theta)|^2\,. \tag{30}$$

The independence of φ in this context arises from the fact that the potential exhibits spherical symmetry. Additionally, the angle θ is measured with respect to the direction k of the incident beam (Firmin et al., 1990).

The total elastic cross section is calculated by performing an integration of σ(θ) across all possible directions, as described in Eqn. (2). Due to the orthogonality of the Legendre polynomials, there are no terms that involve various values of B. Thus:

$$\sigma = \frac{4\pi}{k^2} \sum_{\ell} (2\ell + 1)[\sin(\delta_\ell)]^2 . \tag{31}$$

7.2.2. Calculation of Phase Shifts

The phase shift δA can be determined by solving Eqn. (23) for the radial function R_l (r) in the area where r ≤ a, and then matching it to the form given by Eqn. (24) at r = a. The scattering potential U (r) is non-zero for r < a but becomes zero for r > a. Eqn. (23) has two solutions that are linearly independent. However, only one of these solutions remains finite when r = 0. Therefore, the solution for R_l (r) within the interval 0 < r ≤ a_n is unique, except for normalization. If needed, it can be obtained by numerical methods (Munn et al., 1964). While a boundary condition at r = a requires both RA and to be continuous, it is satisfactory to enforce continuity on the logarithmic derivative;

$$\gamma_\ell = \frac{d \log(R_\ell)}{dr} = \frac{1}{R_\ell}\frac{dR_\ell}{dr}, \tag{32}$$

This gives the following condition:

$$\gamma_\ell = \frac{k[\cos(\delta_\ell)\, j'_\ell(ka) - \sin(\delta_\ell)\, n'_\ell(ka)]}{\cos(\delta_\ell)\, j_\ell(ka) - \sin(\delta_\ell)\, n_\ell(ka)}, \tag{33}$$

A prime symbol represents the differentiation of a function in relation to its argument, and γ_l now represents the logarithmic derivative calculated from the interior at r = a. The phase shift can be determined using the following equation:

$$\tan(\delta_\ell) = \frac{kj'_\ell(ka) - \gamma_\ell\, j_\ell(ka)}{kn'_\ell(ka) - \gamma_\ell\, n_\ell(ka)} . \tag{34}$$

If the scattering potential does not stay uniformly zero at distances greater than a but still exhibits a limited range, it is still feasible to determine shifts in phase as the limit of Eqn. (34) approaches infinity, while keeping in mind that γl is dependent on a. The existence of this limit is contingent upon the potential decreasing at a faster rate than the inverse of r (Yngve & Linnaeus, 1987).

It can be demonstrated that, when B reaches sufficiently high values, the phase shift δA falls in proportion to the reciprocal of the factorial of l. This is an extremely rapid decline, surpassing exponential speed, and it guarantees the convergence of the series Eqn. (29). Nevertheless, this "sufficiently large" value of l might be exceedingly enormous, and the sequence of phase shifts is only useful if it converges within a certain number of terms. In order to determine the circumstances under which this will happen, assume that the potential U (r) is consistently equal to zero. Then, one can express R_l (r) as j_l (kr), where j_l (kr) is directly proportional to kr^l when kr is much less than 1. In this regime, R_1 (r) becomes very small for large values of 1. Now incorporate the potential U (r), which has a range of a, back into Eqn. (23). If the value of ka is a percentage of B, then the function U(r) will be multiplied by the small quantity $R_1(r)$, resulting in a minimal impact on the answer. Based on this somewhat lenient reasoning, one can observe that the phase shift δA will be insignificant as long as the product of the wave number ka is far smaller than 1. Thus, the phase shift series will be particularly valuable when the value of ka is small (Orlov et al., 2017) (Figure 7.3).

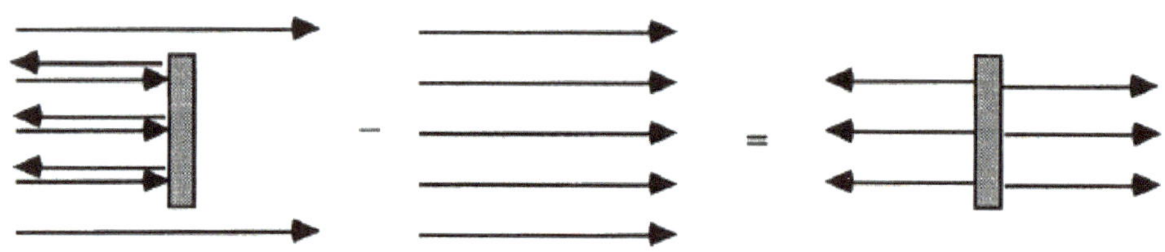

Figure 7.3. Illustration of Flux lines of the total, incident, and scattered waves for a rigid scatterer.

Source: Leslie E, Creative Commons License.

7.3. GENERAL SCATTERING THEORY

The preceding section exclusively discussed scattering caused by a central potential. This type of scattering does not alter the internal states of the particles and hence results in solely elastic scattering. Now examine a broader relationship between both particles. Other factors, like as spin, orientation, and internal states of the particles, may influence the outcome, in addition to the separation distance (Lax & Phillips, 1964).

Collision occurrences can be categorized into multiple types:

a. Elastic scattering, represented as A+ X → A+ X, where the internal states of the particles remain unaltered.
b. Inelastic scattering refers to a process when particle A collides with another particle X, resulting in the production of particles A' and X.' This process is characterized by a change in the internal states of the particles involved (Feder et al., 1983).
c. Rearrangement collisions include the exchange of matter, where the arriving particles may be different from the outgoing particles. This process can be represented as A+X → B+Y. Examples of reactions include nuclear and chemical reactions (Natoli et al., 1990).
d. Particle production occurs when an initial particle A collides with another particle X, resulting in the creation of particles B, Y, and Z in the final state.

The term "channel" is used to describe each way of dissociation of the system into a series of discrete particles. Instead of a singular differential cross-section, it is now necessary to establish differential cross-sections for every single channel (Lambrecht et al., 2006).

Only the elastic and inelastic collisions (a) and (b) shall be considered in this section. The theory of (c) and (d) poses significant technical challenges and would be excessively lengthy for this chapter. Similar to the previous section, the CM variable is not relevant, and one may focus solely on the Hamiltonian for the relative velocity of the particles, in addition to their internal degrees of freedom (Sanchez, 1977). The structure takes the shape:

$$H = \frac{\hbar^2}{2\mu}\nabla^2 + h_1 + h_2 + V$$
$$= H_0 + V .$$

(35)

Here, h_1 and h_2 represent the Hamiltonian operators corresponding to the internal DOF of particles 1 and 2 respectively, whereas μ denotes the reduced mass The Laplacian operator ∇2 operates on the relative coordinate r (Braun & Shabelski, 1988). Based on the previous part, the only focus is on interactions V that decline faster than r^{-1} at high separations. However, the relationship between V and internal variables like spin is not limited. In order to avoid a complicated notation, it is preferable to keep the descriptions for particles and states as short as possible (Mies, 1980).

$$(h_1 + h_2)w_a = e_a\, w_a \, ,$$

(36)

w_a represents a state vector that describes the internal DOF of both particles, whereas e_a represents the total internal energy of the particles. Often, although not universally, w_a can be expressed as a multiplication of state vectors for each particle. However, this information may not be explicitly stated in the notation (Benoit & Hadziioannou, 1988). The requirement of energy conservation is considered satisfied if the starting and final internal conditions are denoted as w_a and w_b, respectively:

$$E = \frac{\hbar^2\, {k_a}^2}{2\,\mu} + e_a = \frac{\hbar^2\, {k_b}^2}{2\mu} + e_b \, .$$

(37)

An elastic collision occurs when the internal states w_a and w_b are identical. If the internal states exhibit disparity, the system will be deemed inelastic, irrespective of the absence of alterations in the internal energy and kinetic energy. The terms representing kinetic energy in Eqn. (37) correspond to the kinetic energies associated with the relative motion in the center of mass frame (Benoit & Hadziioannou, 1988).

7.3.1. Scattering Cross Sections

Schrodinger equation's steady state solutions,

$$H\,\Psi^{(+)}_{\mathrm{a}} = E\,\Psi^{(+)}_{\mathrm{a}} \, .$$

(38)

The label "a" is a combination of both internal and external motion. State labels , where "ℏk/μ" represents the velocity of the incoming particle related to the target, and "a" represents the internal state label. The solution to Eqn. (38) must adhere to an asymptotic boundary condition similar to Eqn. (14). However, in this case, instead of just one outgoing scattered wave, one might have an outgoing wave with separate components corresponding to each potential internal state that could arise from the collision (Sneep & Ubachs, 2005). Thus, the state function will be expressed as:

$$\Psi_{\mathrm{a}}^{(+)} = \sum_{b'} \psi_{\mathrm{a},b'}{}^{(+)}(\mathbf{r})\, w_{b'} \,,$$

(39)

The wave functions exhibit a subsequent asymptotic pattern as the distance r becomes significant:

$$\psi_{\mathrm{a},a}{}^{(+)}(\mathbf{r}) \sim A \left\{ e^{i\mathbf{k}_a \cdot \mathbf{r}} + f_{\mathrm{aa}}{}^{(+)}(\Omega_r)\, \frac{e^{ik_a r}}{r} \right\},$$

(40)

$$\psi_{\mathrm{a},b}{}^{(+)}(\mathbf{r}) \sim A\, f_{\mathrm{a}b}{}^{(+)}(\Omega_r)\, \frac{e^{ik_b r}}{r}, \quad (b \neq a)\,.$$

Ωr represents the angles of vector r. The initial expression, which corresponds to the elastic scattering channel, includes the incoming beam and a wave that is scattered outward. The second statement (Tougaard, 1997), which pertains to inelastic scattering, solely includes an outgoing wave due to the absence of an incident beam associated with the internal state w_b where b is not equal to a

If a = b then Eqn. (41) reduces to the elastic scattering cross section Eqn. (15).

$$\sigma_{a \to b}(\Omega) = \frac{k_b}{k_a} \left| f_{\mathrm{a}b}{}^{(+)}(\Omega) \right|^2$$

(41)

7.3.2. Scattering Amplitude Theorem

The scattering cross sections are defined by the scattering interaction at small distances, even if they only depend on the asymptotic limit of the state function at vast distances. This is achieved through the angular functions (Ω) known as scattering amplitudes, which determine the values of those amplitudes. Next, a theorem will be established that shows a relationship between the asymptotic limit of the scattering state function and an integral that involves the interaction (Mastrolia, 2009).

Did you know?

Scattering theory played a crucial role in the development of quantum mechanics. The famous Rutherford scattering experiment in 1911 provided evidence for the nuclear model of the atom, which led to the development of quantum theory.

proceed with the derivation, it is necessary to propose additional functions, denoted as that resemble scattering functions and are eigenfunctions of H,

$$H\,\Psi_{\mathrm{b}}{}^{(-)} = E\,\Psi_{\mathrm{b}}{}^{(-)}, \tag{42}$$

Satisfying the different asymptotic boundary conditions from Eqn. (40):

$$\Psi_{\mathrm{b}}{}^{(-)} = \sum_{a'} \psi_{\mathrm{b},a'}{}^{(-)}(\mathbf{r})\,\mathbf{w}_{a'}\,, \tag{43}$$

With

$$\psi_{\mathrm{b},b}{}^{(-)}(\mathbf{r}) \sim A\left\{e^{i\mathbf{k}_b\cdot\mathbf{r}} + f_{\mathrm{b}b}{}^{(-)}(\Omega_r)\,\frac{e^{-ik_b r}}{r}\right\} \tag{44}$$

$$\psi_{\mathrm{b},a}{}^{(-)}(\mathbf{r}) \sim A\,f_{\mathrm{b}a}{}^{(-)}(\Omega_r)\,\frac{e^{-ik_a r}}{r}, \quad (a \neq b)\,.$$

These new functionalities comprise a primary beam along with incoming spherical waves. While these states may not align with any probable experimental outcomes, they serve a crucial mathematical function inside the theory (McLoughlin et al., 2022).

The two sets of functions, and , collectively encompass the subspace consisting of positive energy eigenfunctions of H. If the strength of the interaction V is sufficient to generate bound states of the two particles, then such bound states must be included in both sets to ensure their completeness. The presence of two distinct sets of eigenfunctions of H can be comprehended by considering that an eigenvector issue is entirely governed by an operator in addition to boundary conditions. Therefore, Eqn. (38) combined with boundary condition Eqn. (40) represents one well-defined problem, whereas Eqn. (42) with boundary condition Eqn. (44) represents another. Each problem possesses its own distinct set of eigenfunctions (Cheung et al., 2018).

Some writers attempt to circumvent the inherent non-normalizability of the scattering functions by postulating that the cosmos is a vast cube, governed by periodic boundary conditions, where the length of its sides can be infinitely extended in the last stages of the calculation. Every eigenfunction of H can be normalized. However, when periodic boundary conditions are applied to a limited area, the wave solutions described in Eqns. (38) and (42) cannot exist. Consequently, all eigenfunctions of H must be standing waves. While users of the "box normalization" methodology rarely obtain incorrect results, the method itself is inherently inconsistent from the beginning (Stefanov, 2002).

Operator ∇^2 is Hermitian only when applied to functions that meet specific boundary constraints, which are not fulfilled by the scattering state functions. This circumstance is not unusual; in fact, the generalized spectral theorem applies to operators such as

∇^2 that are self-adjoint in Hilbert space but have eigenvectors belonging to a larger space. This is the typical scenario for operators that possess a continuous spectrum of eigenvalues. This information will be essential in developing the scattering amplitude theorem (Martin, 1963).

In order to establish the theorem, analyze two Hamiltonians, each expressed in the form Eqn. (35), but featuring distinct scattering potentials. The equation $H = H_0 + V$, while the equation $H' = H_0 + V$.' Investigate a wave function that represents an outgoing wave eigenfunction of the Hamiltonian H and an incoming wave eigenfunction of the Hamiltonian H,' both corresponding to the same energy level (Khuri & Kinoshita, 1965):

$$H\,\Psi_{\mathrm{a}}{}^{(+)} = E\,\Psi_{\mathrm{a}}{}^{(+)}, \quad H'\,\Psi'_{\mathrm{b}}{}^{(-)} = E\,\Psi'_{\mathrm{b}}{}^{(-)}. \tag{45}$$

The wave vectors are and .

Eqn. (45) can be rewritten as:

$$\frac{\hbar^2}{2\mu}\nabla^2\,\Psi_{\mathrm{a}}{}^{(+)} = (h_1 + h_2 + V - E)\Psi_{\mathrm{a}}{}^{(+)},$$

$$\frac{\hbar^2}{2\mu}\nabla^2\,\Psi'_{\mathrm{b}}{}^{(-)} = (h_1 + h_2 + V' - E)\Psi'_{\mathrm{b}}{}^{(-)}.$$

$$\frac{\hbar^2}{2\mu}\left\{\langle\,\Psi'_{\mathrm{b}}{}^{(-)}|\nabla^2\,\Psi_{\mathrm{a}}{}^{(+)}\,\rangle_R - \langle\,\nabla^2\,\Psi'_{\mathrm{b}}{}^{(-)}|\,\Psi_{\mathrm{a}}{}^{(+)}\,\rangle_R\right\}$$
$$= \langle\,\Psi'_{\mathrm{b}}{}^{(-)}\,|(V - V')|\,\Psi_{\mathrm{a}}{}^{(+)}\,\rangle_R. \tag{46}$$

Or,

If operator ∇^2 were Hermitian in the space of scattering functions, the left-hand side of this equation would disappear. However, this is not the situation. To assess it, employ Eqns. (39) and (43):

$$\frac{\hbar^2}{2\mu}\left\{\langle\,\Psi'_{\mathrm{b}}{}^{(-)}|\nabla^2\,\Psi_{\mathrm{a}}{}^{(+)}\,\rangle_R - \langle\,\nabla^2\,\Psi'_{\mathrm{b}}{}^{(-)}|\Psi_{\mathrm{a}}{}^{(+)}\,\rangle_R\right\}$$
$$= \frac{\hbar^2}{2\mu}\sum_{a'}\sum_{b'}\int_{r\le R}\left\{[\psi'_{\mathrm{b},a'}{}^{(-)}]^*\,\nabla^2\,\psi_{\mathrm{a},b'}{}^{(+)}\right.$$
$$\left. - [\nabla^2\,\psi'_{\mathrm{b},a'}{}^{(-)}]^*\,\psi_{\mathrm{a},b'}{}^{(+)}\right\}\,d^3\mathbf{r}\,\langle\,w_{a'}|w_{b'}\,\rangle. \tag{47}$$

Due to the orthogonality of the internal state vectors, it is possible to simplify the double sum into a single sum using the dummy variable c (Atkinson et al., 1972). The conversion of volume integrals into surface integrals can be achieved through the use of the divergence theorem and the utilization of the identity . This produces:

$$\frac{\hbar^2}{2\mu}\sum_c \oiint_{r=R} \left\{[\psi'_{\mathrm{b,c}}{}^{(-)}]^* \,\nabla\, \psi_{\mathrm{a,c}}{}^{(+)} - [\nabla\, \psi'_{\mathrm{b,c}}{}^{(-)}]^* \,\psi_{\mathrm{a,c}}{}^{(+)}\right\}\cdot d\mathbf{S}\,. \tag{48}$$

Increase the radius R of the sphere to a large enough value so that the wave functions may be replaced with the asymptotic forms, Eqns. (40) and (44). There will be three categories of terms: those that include two spherical waves, those that involve two plane waves, and those that involve a plane wave and a spherical wave. With the exception of a consistent multiplier, the terms in the first category produce (Mastrolia et al., 2012):

$$\sum_c \oiint [f'_{\mathrm{bc}}{}^{(-)}(\Omega)]^*\, f_{\mathrm{ac}}{}^{(+)}(\Omega)\, d\Omega\;\, 2ik_c\;\, e^{i2k_cR}\,.$$

The integration is performed across all orientations on the spherical surface. As the value of R approaches infinity, the exponential function will oscillate extremely quickly with respect to k_c, and can be considered to average down to zero. This can be rationalized by noting that every physical condition inherently encompasses a range of energies, no matter how limited it may be (Hang & He, 2022). Consequently, it is always imperative to integrate over a small range of k_c, which will encompass an endless number of oscillations as the limit R approaches infinity. An analogous reasoning can be applied to remove the expression that includes the interaction of two plane waves. Alternatively, one can convert the surface integral back into a volume integral by reverting from Eqn. (48) to Eqn. (47) for this term, and utilize the orthogonality of plane waves as R approaches infinity. Ultimately, it is necessary to assess the phrases that involve a plane wave and a spherical wave, which are found to have a non-zero value (Alber & Ramm, 1986). The plane wave component of results in the integrals:

$$\oiint_{r=R} [e^{i\mathbf{k}_b\cdot\mathbf{r}}]^*\, f_{\mathrm{ab}}{}^{(+)}(\Omega_r)\left[\frac{\partial}{\partial r}\frac{e^{ik_br}}{r}\right] r^2\, d\Omega_r - \oiint_{r=R}\left[\frac{\partial}{\partial r}\, e^{i\mathbf{k}_b\cdot\mathbf{r}}\right]^* f_{\mathrm{ab}}{}^{(+)}(\Omega_r)\,\frac{e^{ik_br}}{r}\, r^2\, d\Omega_r\,. \tag{49}$$

To assess them, employ the spherical harmonic expansion of a plane wave, which is synonymous with Eqn. (21):

$$e^{i\mathbf{k}\cdot\mathbf{r}} = 4\pi \sum_\ell \sum_m [Y_\ell{}^m(\Omega_r)]^*\, Y_\ell{}^m(\Omega_k)\, i^\ell\, j_\ell(kr)\,. \tag{50}$$

Here, Ωk represents the angle of k, while Ωr represents the angle of r. When the value of r approaches infinity, one can utilize the asymptotic form Eqn. (25) (Degiorgio et al., 2009),

$$j_\ell(kr) \sim \frac{\sin(kr - \frac{1}{2}\pi l)}{kr} = \frac{1}{2ikr}(i^{-l}e^{ikr} - i^{\ell}e^{-ikr}) \tag{51}$$

The asymptotic expansion for r → ∞:

$$e^{i\mathbf{k}\cdot\mathbf{r}} \sim \frac{2\pi}{ikr}\sum_{\ell}\sum_{m}[Y_\ell{}^m(\Omega_r)]^* Y_\ell{}^m(\Omega_k)[e^{ikr} - (-1)^\ell e^{-ikr}]. \tag{52}$$

Substitute only the terms of Eqn. (52) into Eqn. (49),

$$-4\pi\sum_{\ell}\sum_{m}[f_{\mathrm{ab}}{}^{(+)}]_{\ell m}\, Y_\ell{}^m(\Omega_{kb}) + O(R^{-1}) \tag{53}$$

Where,

$$[f_{\mathrm{ab}}{}^{(+)}]_{\ell m} = \oiint [Y_\ell{}^m(\Omega_r)]^*\, f_{\mathrm{ab}}{}^{(+)}(\Omega_r)\, d\Omega_r\,. \tag{54}$$

However, this is merely a factor in the spherical harmonic expansion of the scattering amplitude,

$$f_{\mathrm{ab}}{}^{(+)}(\Omega_{kb}) = \sum_{\ell}\sum_{m}[f_{\mathrm{ab}}{}^{(+)}]_{\ell m}\, Y_\ell{}^m(\Omega_{kb}) \tag{55}$$

By combining these results, including the constant factors that were excluded for the sake of brevity (Bjerrum et al., 2022), and evaluating the limit as R approaches infinity in Eqn. (46), one ultimately obtains:

$$\frac{2\pi\hbar^2}{\mu}|A|^2\left\{[f'_{\mathrm{ba}}{}^{(-)}(-\Omega_{ka})]^* - f_{\mathrm{ab}}{}^{(+)}(\Omega_{kb})\right\} = \langle\, \Psi'_{\mathrm{b}}{}^{(-)}|(V - V')|\Psi_{\mathrm{a}}{}^{(+)}\,\rangle, \tag{56}$$

Is the scattering amplitude theorem. The right-hand side is finite because both V and V' approach zero more rapidly than the reciprocal of r as it tends to infinity. The normalization factor A, as specified in Eqns. (42) and (46), is arbitrary. However, it does not have any impact on Eqn. (56) because it is already implicitly included as a factor in the scattering state functions on the right-hand side (Norris, 1985). A frequent, though not universally adopted, option is to assign the value of A = 1.

The theorem Eqn. (56) has numerous practical applications. For its initial implementation, V = V' and hence ' = f. Consequently:

$$[f_{\mathrm{ba}}{}^{(-)}(-\Omega_{ka})]^* = f_{\mathrm{a}b}{}^{(+)}(\Omega_{kb})\,.$$

(57)

This outcome will be utilized to eliminate the amplitude f $^{-}$, which represents an incoming spherical wave state that is not physically realistic, in support of the more instinctively significant scattering amplitude f $^{+}$. With this comprehension, one can simplify the notation by denoting For the second application, set V' = 0, resulting in H' = H_0 and H = H_0 + V. The eigenfunctions in the lack of any scattering potential are provided by:

$$H_0\,\Phi_{\mathrm{b}} = E\,\Phi_{\mathrm{b}}\,, \quad \Phi_{\mathrm{b}} = A\,e^{i\mathbf{k}_b\cdot\mathbf{r}}\,w_b\,.$$

(58)

Eqn. (55), with A = 1, gives,

$$f_{\mathrm{a}b}{}^{(+)}(\Omega_{kb}) = -\frac{\mu}{2\pi\hbar^2}\,\langle\,\Phi_{\mathrm{b}}|V|\Psi_{\mathrm{a}}{}^{(+)}\,\rangle\,.$$

(59)

While Ω_{kb} (the angles of k_b) is mentioned as a variable, the amplitude also implicitly relies on the fixed direction of $\mathbf{k}_a$ (the direction of the incident beam), as denoted by the subscript a = (k_a, a). For a spherically symmetric contact, the amplitude will solely rely on the relative direction of k_a and k_b. This significant equation represents the scattering amplitude as an integral involving the scattering potential and the state function. This conclusion is noteworthy as it establishes a connection between the behavior of the state function at a significant distance Eqn. (42) and the scattering (Yajima, 1977).

7.4. BORN APPROXIMATION AND DWBA

Two valuable approximations can be deduced from the scattering amplitude theorem discussed in the previous section. The first approximation, known as the Born approximation, is obtained by recognizing that when the scattering potential is weak, the discrepancy between the operators H_0 and $H = H_0 + V$ is minimal (Gambhir & Griffin, 1973). Consequently, the disparity between their eigenfunctions should also be minimal (Hubbard et al., 1983). Therefore, by substituting Eqn. (59), one may derive an estimate for the scattering amplitude by making a substitution with :

$$f_{ab}(\Omega_{kb}) \approx -\frac{\mu}{2\pi\hbar^2} \langle \Phi_b | V | \Phi_a \rangle \,, \tag{60}$$

The second approximation, known as the distorted wave Born approximation (DWBA), is applicable when the scattering potential may be expressed as $V = V_1 + V_2$, with V_2 being of negligible magnitude (Newman et al., 1981). The theorem Eqn. (56) will be utilized to analyze the Hamiltonians $H = H_0 + V_1$ and $H' = H + V_2$. The number Eqn. (56) transforms into:

$$\frac{2\pi\hbar^2}{\mu} \left\{ \left[f'_{ba}{}^{(-)}(-\Omega_{ka}) \right]^* - f_{ab}{}^{(+)}(\Omega_{kb}) \right\} = -\langle \Psi'_b{}^{(-)} | V_2 | \Psi_a{}^{(+)} \rangle \tag{61}$$

However, the Eqn. (57) can be utilized to convert the initial scattering amplitudes (Tamura et al., 1971), resulting in:

$$f'_{ab}{}^{(+)}(\Omega_{kb}) - f_{ab}{}^{(+)}(\Omega_{kb}) = -\frac{\mu}{2\pi\hbar^2} \langle \Psi'_b{}^{(-)} | V_2 | \Psi_a{}^{(+)} \rangle \tag{62}$$

This expression provides an accurate representation of the change in the scattering amplitude caused by the additional scattering potential V_2. If the value of V_2 is of a small magnitude (Demer & Conti, 2003):

$$f'_{ab}{}^{(+)}(\Omega_{kb}) \approx f_{ab}{}^{(+)}(\Omega_{kb}) - \frac{\mu}{2\pi\hbar^2} \langle \Psi_b{}^{(-)} | V_2 | \Psi_a{}^{(+)} \rangle \,, \tag{63}$$

The quantities on the right-hand side of the equation solely pertain to the Hamiltonian (Mytnichenko, 2005) $H = H_0 + V_1$.

ACTIVITY 7.1

Objective: To visualize the scattering of particles and understand the effects of different potentials on scattering patterns.

Materials Needed:

- Computer with internet access
- Simulation software or online applet for quantum scattering (e.g., PhET Interactive Simulations)

a. Procedure:

  - Access a quantum scattering simulation tool.
  - Set up the simulation to observe scattering of particles by a spherical potential.
  - Vary the parameters such as the energy of the incident particles, the range and strength of the potential, and observe the changes in the scattering pattern.
  - Record the differential cross section for different angles and compare with theoretical predictions.
  - Analyze the effect of phase shifts on the scattering amplitude by adjusting the potential and noting the changes in the scattering pattern.

a. Discussion:

  - Discuss how the simulation results compare with theoretical expectations.
  - Explore how changes in the potential affect the phase shifts and the resulting cross sections.
  - Reflect on the importance of the center-of-mass frame in simplifying the analysis of scattering events.

SUMMARY

- This chapter provides a detailed overview of the fundamental concepts and practical applications of scattering theory within the realm of quantum mechanics. Essential concepts, such as the differential and total cross sections, which are used to estimate the probability of scattering occurrences, are developed. Additionally, the significance of the laboratory and center-of-mass frames of reference in the investigation of scattering processes is investigated throughout this chapter.
- By solving the Schrodinger equation under clearly defined boundary conditions, one can conduct an analysis of the scattering phenomenon that occurs in quantum mechanics studies. The chapter provides an explanation of the concept of phase shifts and how they have an effect on the scattering amplitude, which in turn has an effect on the angular distribution of particles that are distributed into the environment. In addition, the chapter introduces the idea of partial wave analysis, which is a technique for performing research on the scattering of particles by spherical potentials. Through the utilization of mathematical derivations and clear explanations, this chapter provides a comprehensive analysis of scattering theory. The purpose of this chapter is to assist readers in comprehending and predicting the interactions between particles.

REVIEW QUESTIONS

1. Explain the difference between the laboratory frame and the center-of-mass frame in the context of scattering experiments.
2. Derive the expression for the differential cross section and explain its significance.
3. Discuss the role of the quantum state function in scattering theory and how it is related to measurable quantities.
4. What are phase shifts, and how do they influence the scattering amplitude?
5. How can the partial wave analysis be used to determine the cross sections for scattering by a spherical potential?
6. Describe the boundary conditions that must be imposed on the Schrödinger equation for scattering problems.

MULTIPLE CHOICE QUESTIONS

1. What does the differential cross section describe?

a. The energy distribution of scattered particles.

b. The number of particles scattered per unit time.

c. The angular distribution of scattered particles.

d. The momentum of the scattered particles.

2. **In the center-of-mass frame, what is conserved during an elastic collision?**
   a. Total momentum
   b. Total energy
   c. Kinetic energy and momentum
   d. Internal energy
3. **What is the significance of the phase shift in scattering theory?**
   a. It determines the speed of scattered particles.
   b. It affects the angular distribution of scattered particles.
   c. It represents the energy change in the system.
   d. It influences the probability density of particle positions.
4. **Which frame of reference is most convenient for theoretical analysis of scattering?**
   a. Center-of-mass frame
   b. Laboratory frame
   c. Inertial frame
   d. Non-inertial frame
5. **The Hamiltonian for a system of two interacting particles in scattering theory includes terms for:**
   a. Potential energy and kinetic energy of individual particles
   b. Kinetic energy of the center of mass and relative motion
   c. Potential energy of the target and projectile
   d. Kinetic energy of individual particles and their interaction potential
6. **What boundary condition must be satisfied by the solution of the Schrödinger equation in scattering theory?**
   a. The wave function must be zero at infinity.
   b. The wave function must be finite at all points.
   c. The wave function must be continuous at the origin.
   d. The wave function must match the incident and scattered flux.

Answers to Multiple Questions

1. (c); 2. (c); 3. (b); 4. (a); 5. (b); 6. (d)

REFERENCES

1. Alba, D., Lusanna, L., & Pauri, M. (2002). Centers of mass and rotational kinematics for the relativistic N-body problem in the rest-frame instant form. Journal of Mathematical Physics, 43(4), 1677–1727.
2. Alber, H. D., & Ramm, A. G. (1986). Scattering amplitude and algorithm for solving the inverse scattering problem for a class of non-convex obstacles. Journal of

Mathematical Analysis and Applications, 117(2), 570–597.

3. Atkinson, D., Johnson, P. W., & Warnock, R. L. (1972). Determination of the scattering amplitude from the differential cross-section and unitarity. Communications in Mathematical Physics, 28(2), 133–158.
4. Benoit, H., & Hadziioannou, G. (1988). Scattering theory and properties of block copolymers with various architectures in the homogeneous bulk state. Macromolecules, 21(5), 1449–1464.
5. Bjerrum-Bohr, N. E. J., Damgaard, P. H., Plante, L., & Vanhove, P. (2022). The SAGEX review on scattering amplitudes Chapter 13: Post-Minkowskian expansion from scattering amplitudes. Journal of Physics A: Mathematical and Theoretical, 55(44), 443014.
6. Blum, A. S. (2017). The state is not abolished, it withers away: How quantum field theory became a theory of scattering. Studies in History and Philosophy of Science Part B: Studies in History and Philosophy of Modern Physics, 60, 46–80.
7. Bobbert, P. A., & Vlieger, J. (1986). Light scattering by a sphere on a substrate. Physica A: Statistical Mechanics and its Applications, 137(1–2), 209–242.
8. Braun, V. M., & Shabelski, Y. M. (1988). Multiple scattering theory for inelastic processes. International Journal of Modern Physics A, 3(11), 2417–2501.
9. Carlson, A. D., Pronyaev, V. G., Smith, D. L., Larson, N. M., Chen, Z., Hale, G. M., & Tagesen, S. (2009). International evaluation of neutron cross section standards. Nuclear Data Sheets, 110(12), 3215–3324.
10. Carpentier, J., Benallegue, M., Mansard, N., & Laumond, J. P. (2016). Center-of-mass estimation for a polyarticulated system in contact—A spectral approach. IEEE Transactions on Robotics, 32(4), 810–822.
11. Charalambopoulos, A., & Dassios, G. (1999). Scattering of a spherical wave by a small ellipsoid. IMA Journal of Applied Mathematics, 62(2), 117–136.
12. Cheung, C., Shen, C. H., & Wen, C. (2018). Unifying relations for scattering amplitudes. Journal of High Energy Physics, 2018(2), 1–28.
13. Degiorgio, V., Potenza, M. A. C., & Giglio, M. (2009). Scattering from anisotropic particles: A challenge for the optical theorem? The European Physical Journal E, 29, 379–382.
14. Demer, D. A., & Conti, S. G. (2003). Reconciling theoretical versus empirical target strengths of krill: Effects of phase variability on the distorted-wave Born approximation. ICES Journal of Marine Science, 60(2), 429–434.
15. Enss, V. (1983). Propagation properties of quantum scattering states. Journal of Functional Analysis, 52(2), 219–251.
16. Feder, R., Rosicky, F., & Ackermann, B. (1983). Relativistic multiple scattering theory of electrons by ferromagnets. Zeitschrift für Physik B Condensed Matter, 52(1), 31–36.
17. Firmin, D. N., Nayler, G. L., Kilner, P. J., & Longmore, D. B. (1990). The application of phase shifts in NMR for flow measurement. Magnetic Resonance in Medicine, 14(2), 230–241.

18. Fushman, I., Englund, D., Faraon, A., Stoltz, N., Petroff, P., & Vuckovic, J. (2008). Controlled phase shifts with a single quantum dot. Science, 320(5877), 769–772.
19. Gambhir, B. L., & Griffin, J. J. (1973). Rigorous solutions for three-body rearrangement reactions and validity of the distorted-wave Born approximation. Physical Review C, 7(3), 1006.
20. Hang, Y. F., & He, H. J. (2022). Structure of Kaluza-Klein graviton scattering amplitudes from the gravitational equivalence theorem and double copy. Physical Review D, 105(8), 084005.
21. Hibino, K., Oreb, B. F., Farrant, D. I., & Larkin, K. G. (1997). Phase-shifting algorithms for nonlinear and spatially nonuniform phase shifts. JOSA A, 14(4), 918–930.
22. Hubbard, L. M., Shi, S. H., & Miller, W. H. (1983). Multichannel distorted wave Born approximation for reactive scattering. The Journal of Chemical Physics, 78(5), 2381–2387.
23. Hughes, T. P., Rodrigues, M. J., Bellwood, D. R., Ceccarelli, D., Hoegh-Guldberg, O., McCook, L., & Willis, B. (2007). Phase shifts, herbivory, and the resilience of coral reefs to climate change. Current Biology, 17(4), 360–365.
24. Hunziker, W., & Sigal, I. M. (2000). Time-dependent scattering theory of N-body quantum systems. Reviews in Mathematical Physics, 12(08), 1033–1084.
25. Kesmodel, U. S. (2018). Cross-sectional studies—What are they good for? Acta Obstetricia et Gynecologica Scandinavica, 97(4), 388–393.
26. Khuri, N. N., & Kinoshita, T. (1965). Real part of the scattering amplitude and the behavior of the total cross section at high energies. Physical Review, 137(3B), B720.
27. Koch, H. W., & Motz, J. W. (1959). Bremsstrahlung cross-section formulas and related data. Reviews of Modern Physics, 31(4), 920.
28. Kriman, A. M., Kluksdahl, N. C., & Ferry, D. K. (1987). Scattering states and distribution functions for microstructures. Physical Review B, 36(11), 5953.
29. Lambrecht, A., Neto, P. A. M., & Reynaud, S. (2006). The Casimir effect within scattering theory. New Journal of Physics, 8(10), 243.
30. Lax, P. D., & Phillips, R. S. (1964). Scattering theory. Bulletin of the American Mathematical Society, 70(1), 130–142.
31. Lax, P. D., & Phillips, R. S. (1971). Scattering theory. The Rocky Mountain Journal of Mathematics, 1(1), 173–223.
32. Lundstrom, M. (1997). Elementary scattering theory of the Si MOSFET. IEEE Electron Device Letters, 18(7), 361–363.
33. Martin, A. (1963). Unitarity and high-energy behavior of scattering amplitudes. Physical Review, 129(3), 1432.
34. Martinez-Haya, B., Aoiz, F. J., Banares, L., Honvault, P., & Launay, J. M. (1999). Quantum mechanical and quasiclassical trajectory study of state-to-state differential cross sections for the F + D2 → DF + D reaction in the center-of-mass and laboratory frames. Physical Chemistry Chemical Physics, 1(15), 3415–3427.
35. Mastrolia, P. (2009). Double-cut of scattering amplitudes and Stokes' Theorem.

Physics Letters B, 678(2), 246–249.

36. Mastrolia, P., Mirabella, E., Ossola, G., & Peraro, T. (2012). Scattering amplitudes from multivariate polynomial division. Physics Letters B, 718(1), 173–177.
37. McLoughlin, T., Puhm, A., & Raclariu, A. M. (2022). The SAGEX review on scattering amplitudes chapter 11: Soft theorems and celestial amplitudes. Journal of Physics A: Mathematical and Theoretical, 55(44), 443012.
38. Mekhov, I. B., & Ritsch, H. (2009). Quantum optics with quantum gases: Controlled state reduction by designed light scattering. Physical Review A—Atomic, Molecular, and Optical Physics, 80(1), 013604.
39. Mies, F. H. (1980). A scattering theory of diatomic molecules: General formalism using the channel state representation. Molecular Physics, 41(5), 953–972.
40. Munn, R. J., Mason, E. A., & Smith, F. J. (1964). Some aspects of the quantal and semiclassical calculation of phase shifts and cross sections for molecular scattering and transport. The Journal of Chemical Physics, 41(12), 3978–3988.
41. Mytnichenko, S. V. (2005). Distorted-wave Born approximation in the case of an optical scattering potential. Physica B: Condensed Matter, 355(1–4), 244–249.
42. Natoli, C. R., Benfatto, M., Brouder, C., López, M. R., & Foulis, D. L. (1990). Multichannel multiple-scattering theory with general potentials. Physical Review B, 42(4), 1944.
43. Newman, K. E., Teitel, S., & Domany, E. (1981). The distorted wave Born approximation: Application to elastodynamics. Journal of Applied Physics, 52(7), 4363–4370.
44. Norris, A. N. (1985). Radiation from a point source and scattering theory in a fluid-saturated porous solid. The Journal of the Acoustical Society of America, 77(6), 2012–2023.
45. Orlov, Y. V., Irgaziev, B. F., & Nabi, J. U. (2017). Algorithm for calculations of asymptotic nuclear coefficients using phase-shift data for charged-particle scattering. Physical Review C, 96(2), 025809.
46. Rauch, J., & Taylor, M. (1975). Potential and scattering theory on wildly perturbed domains. Journal of Functional Analysis, 18(1), 27–59.
47. Sanchez, N. (1977). Wave scattering theory and the absorption problem for a black hole. Physical Review D, 16(4), 937.
48. Setterfield, T. P., Miller, D. W., Leonard, J. J., & Saenz-Otero, A. (2018). Mapping and determining the center of mass of a rotating object using a moving observer. The International Journal of Robotics Research, 37(1), 83–103.
49. Sneep, M., & Ubachs, W. (2005). Direct measurement of the Rayleigh scattering cross section in various gases. Journal of Quantitative Spectroscopy and Radiative Transfer, 92(3), 293–310.
50. Stefanov, P. (2002). Estimates on the residue of the scattering amplitude. Asymptotic Analysis, 32(3, 4), 317–334.
51. Strauss, W. A. (1981). Nonlinear scattering theory at low energy. Journal of Functional Analysis, 41(1), 110–133.

52. Tamura, T., Coker, W. R., & Rybicki, F. (1971). Distorted wave Born approximation for nuclear reactions. Computer Physics Communications, 2(2), 94–106.
53. Tannor, D. J., & Weeks, D. E. (1993). Wave packet correlation function formulation of scattering theory: The quantum analog of classical S-matrix theory. The Journal of Chemical Physics, 98(5), 3884–3893.
54. Tougaard, S. (1997). Universality classes of inelastic electron scattering cross-sections. Surface and Interface Analysis: An International Journal Devoted to the Development and Application of Techniques for the Analysis of Surfaces, Interfaces and Thin Films, 25(3), 137–154.
55. Videen, G. (1991). Light scattering from a sphere on or near a surface. JOSA A, 8(3), 483–489.
56. Wang, M., & Marshall, A. G. (1990). Laboratory-frame and rotating-frame ion trajectories in ion cyclotron resonance mass spectrometry. International Journal of Mass Spectrometry and Ion Processes, 100, 323–346.
57. Yajima, K. (1977). Scattering theory for Schrödinger equations with potentials periodic in time. Journal of the Mathematical Society of Japan, 29(4), 729–743.
58. Yngve, S., & Linnaeus, S. (1987). On the calculation of phase shifts produced by complex potentials. III. Physical Review D, 35(12), 3847.
59. Zhang, J., & Ho, T. L. (2018). Potential scattering on a spherical surface. Journal of Physics B: Atomic, Molecular and Optical Physics, 51(11), 115301.
60. Zhang, X., Whelan, C. T., & Walters, H. R. J. (1992). Distorted-wave born approximation calculations of (e, 2 e) reactions. Zeitschrift für Physik D: Atoms, Molecules and Clusters, 23, 301–308.

CHAPTER

8 Application of Quantum Mechanics: Quantum Cryptography

LEARNING OBJECTIVES

After studying this chapter, you will be able to:

1. Understand the basic principles of quantum cryptography
2. Comprehend the concept of a qubit
3. Comprehend the steps involved in the BB84 quantum key distribution protocol
4. Understand techniques such as error correction and privacy amplification used to ensure the integrity and security of the key
5. Comprehend the applications of quantum cryptography in commercial systems

KEY TERMS FROM THIS CHAPTER

authentication
error correction
quantum channel
quantum key distribution
BB84 protocol
polarized light
qubit

UNIT INTRODUCTION

Information carriers that transfer data in microscopic systems such as individual photons or atoms follow the laws of quantum physics rather than those of conventional physics. There are numerous opportunities for information processing that arise from this, as it allows for the creation of innovative information processes that were previously hindered by classical physics (Bennett et al., 1992).

Quantum cryptography is a highly advanced and well-developed technique within the emerging topic of quantum information processing. Quantum cryptography's security is based on physics, as opposed to cryptographic systems that depend on hypothetical mathematical assumptions. It is being developed now with the expectation that in the future, it will be possible to break conventional public-key ciphers (Portmann & Renner, 2022).

One day, a quantum computer might be able to interpret modern codes, for example. The one-time pad is still impenetrable even with such advanced techniques. The one-time pad's flaw is that it requires the message it is meant to encrypt to be safely delivered to the intended recipient for the secret, random, symmetric key to work. Moreover, the key has a single usage limit. Using single quantum particle behavior, quantum cryptography solves this key distribution problem in a way that is not possible with merely conventional physics (Singh, 2022).

Understanding the fundamentals of quantum cryptography can be achieved by exploring the concept of transmitting information through individual photons. In the field of quantum physics, a solitary photon has the remarkable ability to serve as a quantum bit, commonly known as a qubit. To determine the value of a qubit, it is necessary to measure the corresponding property of the photon, such as polarization (Pirandola et al., 2020). Based on the principles of quantum physics, it is inevitable that any measurement will cause a change in the property being observed. This poses a significant challenge for individuals attempting to intercept the transmission, as the alterations caused by the measurement can be readily identified by both the sender and receiver. Due to the nature of the transmission, it is not feasible to utilize this concept for the direct transmission of confidential messages. However, using one-time-pad cryptography, it can send a symmetrical, secret, and unpredictable key. Both the sender and the recipient will quickly identify any attempt at eavesdropping if the transmission is intercepted. When this happens, the key can be quickly thrown away, enabling the sender to send another key until a safe one is obtained (Techateerawat, 2010).

Although quantum cryptography has its roots in fundamental principles, it was not introduced until 1970 when Stephen Wiesner presented it in an unpublished manuscript. Prior to the publication of a groundbreaking work by Charles Bennett and Gilles Brassard in 1984, the subject matter was largely overlooked. Technology for implementing quantum cryptography in real-world systems is currently accessible. The goals of this chapter are to outline the fundamentals of quantum cryptography and provide examples of current technology-based quantum cryptography protocols and implementations (Steane, 1998).

8.1. QUANTUM BIT

Every information can be broken down into elementary units, commonly referred to as bits. In the field of quantum physics, a bit is a binary unit that can be expressed as either a 0 or a 1. Nevertheless, understanding the process of transferring information to a qubit is distinct from transferring information to a classical bit, that will be soon discovered (Chuang & Yamamoto, 1996).

Consider a (qu)bit as a container, as seen in Figure 8.1, where we can insert a ball of one of two colors to hold one of two bit values. One must carefully open the box and note the color of the ball inside in order to ascertain the bit value of the box. When using a classical approach, the ball inside stays the same color as when it was first put in the box (Lo & Chau, 1998).

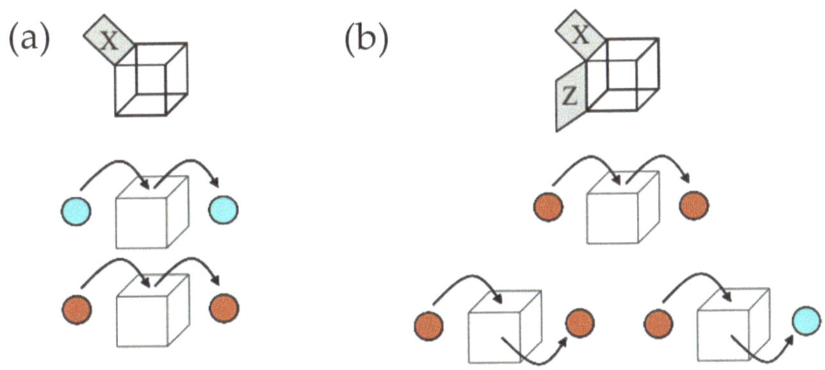

Figure 8.1. Illustration of classical versus quantum bit.

Note: (a) Classical bit: When the ball is placed inside traditional box, ball color that emerges is same as the color of the box. (b) Qubit: The ball will emerge in a random hue if we place it in a quantum box and open the incorrect door.

Source: Lars Lydersen, Creative Commons License.

However, this is not always true for qubits. Within the framework of quantum formalism, the two distinct doors of the box symbolize two distinct methods of measuring the value of the qubit. In order to retrieve accurate bit data, it is essential to determine the specific door that was utilized for storing the qubit and thereafter use the identical door. The information stored in the qubit will be changed to an arbitrary bit value if the wrong door is opened, resulting in an unanticipated color for the ball within. As a result, the data that was recorded is deleted (Grassl et al., 1998).

A polarized photon serves as an example of a qubit. A technique for determining the polarization of the photon involves sending it through a polarizing beamsplitter and noting the specific output from which the photon emerges. However, the polarizing beamsplitter has constraints that prevent it from being oriented at two different angles simultaneously.

Therefore, the qubit value remains unreadable without the necessary supplementary data. For example, the qubit value can be obtained by aligning the beamsplitter with the axis of a given reference coordinate system if one knows that the polarization in that coordinate system is either horizontal or vertical. As seen in Figure 8.2, the detection of a photon at one of the beamsplitter's outputs indicates that the photon's polarization was either horizontal or vertical (Ofek et al., 2016).

In other words, in order to read a qubit correctly, one must know in advance which coordinate system was used to prepare it. The measurement result will be random if the beamsplitter is oriented differently, similar to what happens in Figure 8.1 when the wrong door of the quantum box is opened. The way nature operates at the quantum level is demonstrated by the fact that, once a photon is detected in one of the outputs of the beamsplitter, it really assumes the output polarization and loses all trace of its original polarization (Murch et al., 2013).

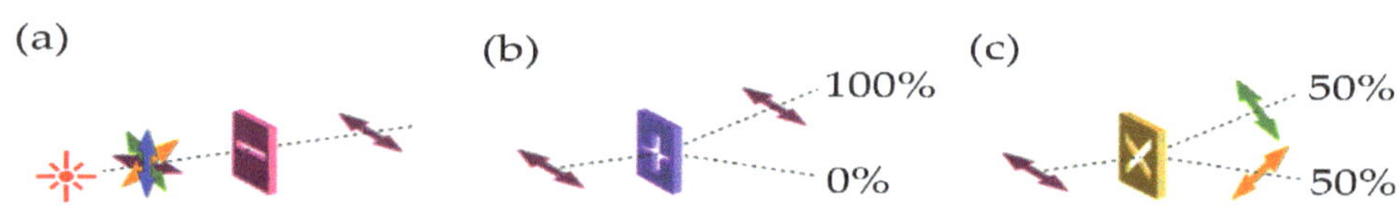

Figure 8.2. Illustration of qubit as a polarized photon.

Source: Lars Lydersen, Creative Commons License.

Did you know?

Quantum cryptography uses aspects of quantum mechanics' unusual and counterintuitive properties, such as entanglement and superposition, to secure communication channels.

Remember

QKD can theoretically provide unconditional security, meaning that its security does not rely on computational assumptions, unlike classical cryptography

8.2. QUANTUM COMPUTING

In order to duplicate a qubit, it is necessary to extract the bit value via accessing the quantum box. Nevertheless, it is impossible to determine the specific door that was utilized to store the bit value of the qubit. Randomly selecting one of the doors could potentially result in the corruption of the data stored in the qubit. Therefore, due to the inherent imperfections in reading quantum bits, it is also impossible to properly replicate them. Typically, the capacity to replicate information is widely seen as highly advantageous (Brassard et al., 1998).

However, this would be devastating in secure communication because the person listening in may record the conversation and save a copy of the message. Qubits, however, cannot be duplicated. Secure communication can be facilitated by utilizing the non-copying characteristic of quantum information. Consequently, qubits can be utilized to transmit a key from sender to recipient without giving an eavesdropper the opportunity to obtain a duplicate covertly (Cao et al., 2019).

8.3. QUANTUM KEY DISTRIBUTION

In Figure 8.3, the process of quantum cryptography is shown, which involves the distribution of a random secret key instead of directly transmitting confidential information. After the successful transmission of the key, it can be used with a traditional symmetric cipher to both encrypt and decrypt data. Let us examine the quantum key distribution protocol (Broadbent & Schaffner, 2016).

8.3.1. The BB84 Protocol

Let's use the names Bob, the recipient, and Alice, the sender, to show the protocol. Assume Alice creates a random sequence of bits, uses door X or door Z of the quantum box to randomly code the bits into qubits, and then delivers the qubits to Bob via a quantum channel (Shor & Preskill, 2000).

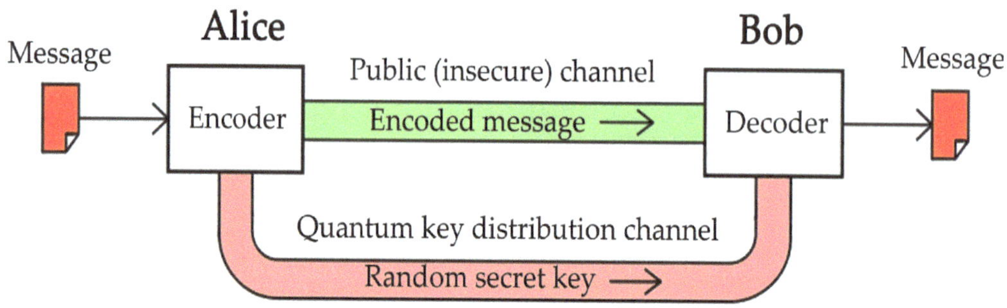

Figure 8.3. Schematic of utilizing symmetric encryption with quantum key distribution. The first stage is giving Bob and Alice the secret key. Then a symmetric cipher using the key can encode and decode data that is transferred.

Source: Vadim Makarov, Creative Commons License.

Bob chooses doors at random since he has no idea which ones Alice used. Bob only opens the correct door half the time as a result. He reads the appropriate information in certain situations. At this point, Bob's pieces are referred to as the raw key. Bob and Alice both make public the doors that were used to store and measure the values of the qubits after they have unlocked all of the quantum boxes. The only qubit values that are saved are those from the boxes that happened to share doors. Both Alice and Bob now have access to the sifted key, which is a shorter version of the original raw key made up of bits chosen at random (Molotkov, 2019). What will happen if Eve, the eavesdropper, tries to access specific quantum boxes during the transmission? In the unlikely situation that Eve chooses the right door, she will be able to replicate the information and send it to Bob. Nevertheless, there is a 50% chance that she may select the incorrect door, potentially altering the qubit's value. Alice and Bob can check the accuracy of Bob's information by conducting a test and comparing a small amount of their key.

They can be certain that Eve did not open any boxes if the important parts of Alice and Bob match. Nevertheless, if the keys do not match, it suggests that Eve made an effort to measure the key. The protocol BB84, which we have just described, was introduced in 1984 by Bennett and Brassard. Considering the measurement of error fractions in the key, commonly referred to as the quantum bit error rate, the protocol offers either a guaranteed secure key or notifies Bob and Alice of a failed key distribution (Lee et al., 2022).

8.3.2. The BB84 Protocol Using Polarized Light

Polarized single photons can be used to carry out the BB84 technique, as shown in Figure 8.4. Alice can use either –45 (bit value 0) or +45 (bit value 1) polarization to code the qubit, or she can use both horizontal (bit value 0) and vertical (bit value 1) polarization. Bob utilizes two photon detectors after the beamsplitter and two interchangeable polarizing beamsplitters to receive the qubits (Kak et al., 2011). Bob can distinguish between the vertical and horizontal polarizations using one polarizing beamsplitter, and between the –45° and +45° polarizations using the other polarizing beamsplitter.

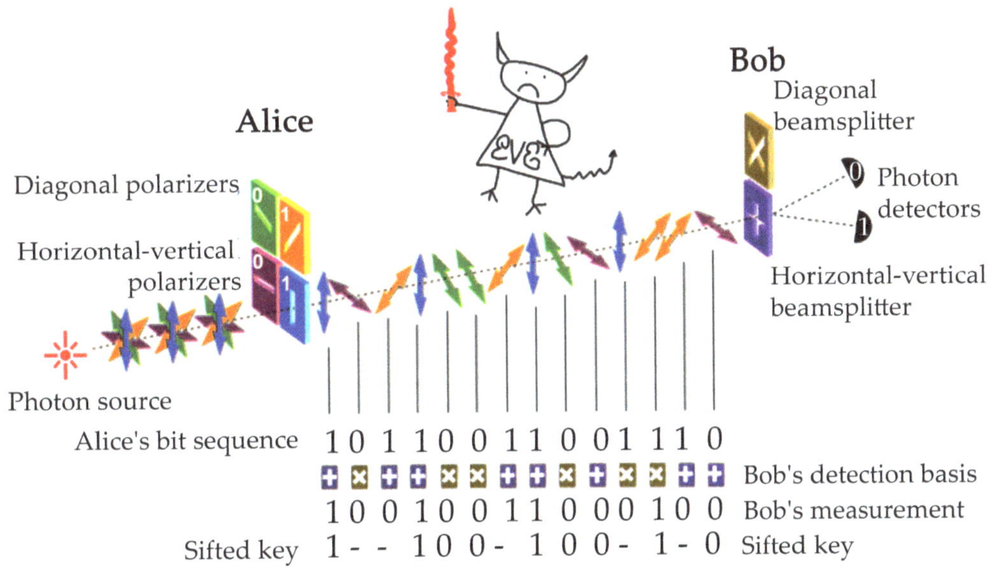

Figure 8.4. Illustration of BB84-protocol using polarized light.

Source: Vadim Makarov, Creative Commons License.

When Bob utilizes a polarizing beamsplitter that aligns completely with Alice's polarization choice, he will be able to accurately determine the state of polarization. This will indicate that he has indeed made the correct selection. If Bob uses a polarizing beamsplitter that is not suitable for Alice's chosen polarization, he will not be able to gather any information about the state of polarization. Simply put, he will have chosen the wrong door (Choe et al., 2018). Once Bob has gathered a sufficient number of photons, which form the raw key, he proceeds to publicly communicate the specific sequence of polarizing beamsplitters he utilized. However, it is crucial to note that he does not disclose the outcome of the measurement. When Bob uses the beamsplitter successfully, Alice tells him about those occasions by comparing this sequence to the bases she used, but noticeably without revealing the polarization she sent (Khaleel & Tawfeeq, 2009).

Alice and Bob are aware that the bits, which make up the filtered key, have identical values, as long as there was no interference from an eavesdropper during the transmission. Alice and Bob assess the security of the message by randomly selecting a portion of the sifting key and comparing it through the public channel. When the transmission is intercepted or disrupted, the correlation between the bit values will diminish, leading to a rise in the quantum bit error rate (Czermann et al., 2021). Every eavesdropping technique causes some sort of systemic disturbance. Consequently, Alice and Bob can be certain that the communication was not intercepted if they do not detect any differences in the subset of the key, and they can encrypt using the remaining portion of the key (Ramos et al., 2022).

8.4. PRACTICAL QUANTUM CRYPTOGRAPHY

Every implementation of quantum key distribution utilizes currently available technology, which implies that the system components, including photon sources, transmission channels, polarizing beamsplitters, and photon detectors, are inherently flawed. This fact carries significant consequences. A universal flaw in all components is the occasional loss of photons (Wang, 2005).

In a practical scenario, the majority of the photons emitted by Alice are likely to be absorbed during transmission, and even if they do reach the detector, they may not always produce a click. In practice, the bits that truly contribute to the raw key are only the photons that have been detected as clicks in Bob's detectors (Zbinden et al., 1998).

Another consequence of defects is that the qubits are not precisely prepared and detected in the theoretical basis. Technological flaws will cause faults in the filtered key, which cannot be differentiated from errors caused by any attempts to eavesdrop. Current technology often exhibits error rates that fall within the range of a few percent. False detection signals, also known as dark counts, can often have a significant impact on the quantum bit error rate. These signals often originate from the photon detectors (Jain et al., 2014).

8.4.1. Error Correction and Privacy Amplification

Alice and Bob are uncertain about whether the flaws in the sifted key were caused by gadget defects or by eavesdropping. They must adopt an optimistic viewpoint and presume that all mistakes were caused by unauthorized listening. In this phase of the protocol, Bob and Alice exchange standard information that is correlated to a high extent, although not perfectly. Additionally, they assume that Eve, the third person, is aware of this information. Classical information theory can help with this problem by providing methods to extract a shorter, error-free key that Eve is not aware of (Tsurumaru, 2021).

Initially, Alice and Bob have to use classical error correcting techniques to get identical keys. Eve possesses further knowledge regarding this key, as Alice and Bob have been forced to disclose more information when engaging in public communication during the error correction process. As a result, the last phase of the quantum cryptography protocol has a privacy amplification mechanism that shrinks the key and limits Eve's access to its information (Maurer & Wolf, 2003).

Bob and Alice conduct privacy amplification by using a randomly chosen hash function from the universal-class to process the error-corrected key. When Bob has a greater understanding of Alice's sifted key compared to Eve, privacy amplification will lead to a final key that is much shorter, rendering Eve's knowledge about it insignificant (Watanabe et al., 2006).

To provide a sense of the numbers, 2000 bits can be reduced to 754 secret bits, about which Eve's knowledge is little (less than 10–6 bit), assuming a realistic quantum bit error rate of 4% and eavesdropping as the primary source of information. From the initial 2000 bits, one can extract 105 secret bits with an 8% quantum bit error rate. Figure 8.5 shows the subsequent sequence of steps in the general quantum key distribution process (Chen et al., 2024).

Did you know?

The BB84 protocol, one of the first quantum key distribution protocols, was developed in 1984 by Gilles Brassard and Charles Bennett. The "BB" in BB84 stands for their initials.

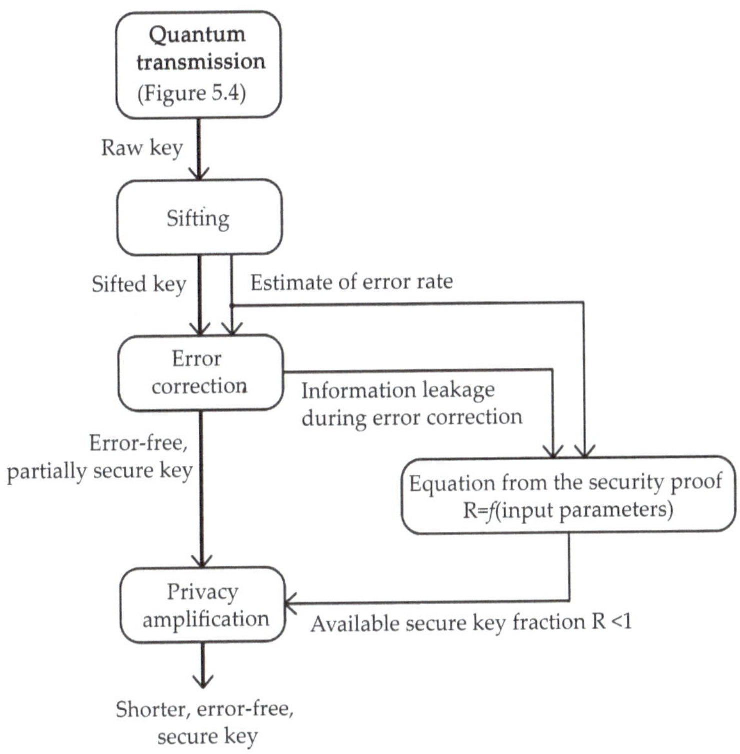

Figure 8.5. Representation of classical post-processing in quantum key distribution.

Source: Dag Roar Hjelme, Creative Commons License.

8.4.2. Security Proofs

It is easy to understand why quantum key distribution yields a fully secret key. On the other hand, the proofs' details are highly complex. It is established that, given a perfect implementation of the protocol and the assumption that Eve can only interact with one qubit at a time, Bob will always know more than Eve, so long as the quantum bit error rate remains below 14.65% (Fischlin & Schnorr, 2000).

It is established that secure transmission requires a quantum bit error rate of less than 11%. This implies that Eve could do whatever made possible by accepted physics even if she had infinite power and could assault an endless number of qubits coherently. The security proof gives an equation to calculate the necessary privacy amplification as long as the mistake rate is less than this cutoff (Figure 8.5) (Duc et al., 2019).

The security has been rigorously demonstrated for specific idealized models of equipment. Nevertheless, a significant portion of the current discussion revolves around the potential existence of imperfections in actual hardware that have not yet been considered in the proofs. The focus lies on finding ways to address and eliminate these loopholes (Bai et al., 2018).

8.4.3. Authentication

One issue still has to be resolved: when Alice and Bob produce a secret key, how can they be certain that they are actually speaking to each other on the public channel and not Eve? Eve might be positioned halfway between Alice and Bob, posing both as Alice and as Bob to Alice. In order to prevent this, it is known that Alice and Bob must begin with a brief common secret (a few 100 bits) in order for them to be able to identify one another on the first run of the protocol (Barkadehi et al., 2018).

They can authenticate in subsequent runs using a portion of the secret key they generate after the first successful key distribution. Quantum key distribution has been shown to provide far more secret keys than they need for authentication. Quantum secret growing protocols are analogous to quantum key distribution in this regard (Lyu et al., 2004).

The requirement for initial authentication is fundamental and universal in all forms of cryptography: without it, there would be no way to confirm that you are communicating with the intended recipient and not an unauthorized entity. All cryptographic methods share the usage of an initial trusted key or biometric authentication, which includes things like verifying a pen signature, chatting with a known individual over the phone, or being present in person throughout the transaction (Velásquez et al., 2018).

8.5. TECHNOLOGY

Quantum cryptography is essentially made possible by two technologies: single photon detectors and single photon sources. Furthermore, a transmission channel, often referred to as a quantum channel, is required for the single photon states. The remainder of the system is implemented with reasonably common electronic and telecommunications technology (Basu et al., 2006).

8.5.1. Single Photon Sources

It is quite challenging to achieve the realization of single photon sources. As a result, the majority of systems in use today depend on subtle laser pulses. There is typically less than one photon per pulse in typical laser pulses, such as those produced by semiconductor lasers, because they are so weak (Somaschi et al., 2016).

An inherent drawback of this approach is the higher probability of having several photons per pulse, unless the average photon count is substantially less than one. The number of photons in a pulse obeys Poisson statistics. When the average photon number is 0.1, the probability of not finding any photons is high (0.9048), the probability of finding one photon is moderate (0.0905), and the probability of finding two or more photons is low (0.0047) (Senellart et al., 2017).

Eve can catch and hold onto one of the photons in a pulse that Alice generates if it has several photons until the basis is discovered. Subsequently, she can execute an accurate measurement in this specific framework, acquiring knowledge of the binary value of the qubit transmitted to Bob (Babinec et al., 2010).

Hence, the existence of multiphoton pulses diminishes the rate at which secret keys are generated. Decreasing the average photon number can reduce the proportion of multiphoton pulses compared to single-photon pulses. However, this also leads to a higher number of vacant bit slots and hence a lower bit rate (Uppu et al., 2020).

It is possible to address this issue by increasing the pulse rate. However, there is still a disadvantage to consider, as the dark counts in single photon detectors are rather large. As the average photon number decreases, the signal to noise ratio takes a hit, resulting in an increase in the quantum bit error rate (Collins et al., 2013).

The ideal photon source is thought to be a device that emits single photons as needed. Despite advancements being made, the development of practical devices in this field is still pending. However, significant progress has been made in achieving practical operation over long distances by utilizing low-intensity laser pulse sources. Furthermore, there are advanced protocols that allow for secure operation over distances greater than 100 km using a faint laser pulse source (Barnes et al., 2002).

8.5.2. Single Photon Detectors

There are various methods available to achieve single photon detection, including the use of photomultipliers, avalanche photodiodes, and even some unique superconducting devices that require cryogenic cooling below 4 K. The avalanche photodiode is currently the most feasible choice for quantum cryptography. Avalanche photodiodes are semiconductor components that are operated under a large voltage in order to detect single photons (Zappa et al., 1996).

A single electron is excited by a single photon that enters the semiconductor and is absorbed. This first electron turns into an avalanche of electrons (several thousands) when it collides with the semiconductor's lattice due to the high electric field present in the device. This avalanche is big enough for an external circuit to detect it as a current pulse (Najafi et al., 2015).

Unfortunately, an avalanche can also happen in the absence of a photon, triggered by thermal stimulation, tunneling, or the release of trapped carriers. The latter occurs as a result of trapped electrons from a previous avalanche being gradually released from semiconductor lattice defects. In practice, the achievable count rate is limited by the release of trapped carriers. One major limitation of the current systems using weak laser pulses is that high pulse rates are required to achieve acceptable bit rates (Komiyama et al., 2000).

8.5.3. Quantum Channel

A strong connection between Alice and Bob is crucial in the field of quantum physics. The channel must be designed in a way that ensures the qubit is shielded from any interference caused by external factors. It has been demonstrated that single photon states (qubits) can be sent very effectively through typical single-mode optical fiber in telecommunication and data applications. Transmission losses in optical fibers can pose a limitation on the number of qubits that reach the detector. The key exchange rate is significantly influenced by the photon transmission probability of the link, which has a direct impact on it (Holevo, 1998).

Telecommunication fibers in the modern era experience transmission losses of around 0.35 dB/km, 2 dB/km, and 0.2 dB/km at the widely utilized communication wavelengths of 800 nm, 1,300 nm, and 1,550 nm, respectively. At a wavelength of 1,550 nm, it may be inferred that a minimum of 50% of the photons are lost at a distance of 15 km, while 99% of the photons are lost at a distance of 100 km. The longest quantum key distribution achieved in controlled laboratory settings thus far is 250 kilometers, with a notably low transmission rate of 15 secret bits per second. Presently, commercial systems have a restricted range of 50–100 kilometers (Devetak, 2005).

Environmental fluctuations, like temperature changes, can affect all types of fibers. This alters the polarization state, consequently modifying the qubit values. As a result, the imperfect channel's presence increases the error rate. This polarization state perturbation's overall effect causes a conversion between the fiber's input and output. In the event that this transformation is stable, Alice and Bob can adapt by aligning their systems with a polarization controller and indicating the direction of the diagonal and vertical polarization, for example (Pirandola et al., 2019).

An active feedback system can be used to maintain constant alignment over time if the transition occurs gradually. It is possible to find smart solutions: Back in the day, commercial systems used an optical method that was quite handy, called "plug and play." With this technique, polarization perturbation was effectively minimized without the need for active control.

Instead of using fiber, a direct passage through the atmosphere can be utilized as the quantum channel. Alice and Bob use little telescopes aimed at each other to convey photons. The presence and standard of such a connection is evidently influenced by meteorological conditions. Nevertheless, air does not disturb polarization nor does it exhibit significant loss. The longest transmission distance recorded is 144 km, accomplished between hilltops on the Canary Islands. However, for practical purposes, networks spanning 10–30 km may be more feasible (DiVincenzo et al., 1998).

The possibility of a secret key being exchanged between a satellite and a ground station is increased by the success of these ground experiments. Thus, a low-orbit satellite can provide a global key distribution network by gradually establishing key distribution links with places below its flight path.

8.5.4. Random Number Generator

The one-time pad key needs to be completely random. Computers cannot be utilized to generate random numbers for cryptography systems since they are deterministic systems. As a result, a physically random mechanism must produce the random numbers. A beamsplitter that receives a single photon as an example can be discovered in one of the device's two exits. According to quantum mechanics, the exit at which it is located is random (Marsaglia et al., 1990).

Undoubtedly, there are a ton of additional methods that are available. Large bit rate real random number generators are still in the experimental stage, despite the fact that commercial quantum key distribution systems currently in use physical random number generators with bitrates of a few Mbit/s (Jennewein et al., 2000).

8.6. APPLICATIONS

Even though quantum cryptography has reached a level of technological maturity, its commercial success remains limited to a small niche market, leaving its potential for wider adoption uncertain. Classical cryptographic systems based on assumptions on computational complexity are highly effective and convenient. Efficient, cost-effective, capable of operating at high speeds across vast distances (Becke et al., 2006).

As stated in the Introduction of this chapter, the security of these approaches is not foolproof against future advancements in cryptanalysis (and even presently, it is not entirely certain). Nevertheless, their convenience is unmatched. If any of them were to fail, this wouldn't be the first example in history where convenience was prioritized over rigorous security measures (Reuss et al., 2005).

However, establishing a quantum key distribution link currently comes with limitations in terms of distance and bitrate, making it a more costly project. Interestingly, utilizing a trusted courier to transport a hard disk containing random numbers could yield a greater supply of keys compared to the lifespan of most quantum key distribution links.

It is important to keep in mind that the connection for quantum key distribution necessitates a short key for initial authentication, resulting in involvement of a trusted courier. Nevertheless, capability for unlimited growth of essential elements from the short authentication key enables quantum cryptography to effectively expand within a network of numerous users. Conversely, in the scenario of distributing data on hard disks, the necessary storage capacity would rapidly become impractical (Mason, 1981).

8.6.1. Commercial Systems with Dual Key Agreement

In modern commercial systems, quantum key distribution serves as an additional layer of security alongside classical key distribution and encryption, as shown in Figure 8.6. It is possible to combine keys sent through public key cryptography with keys received by quantum key distribution. This is accomplished by utilizing the exclusive-OR binary function to encrypt one key while using the second key as a one-time pad (Kumar et al., 2018).

The resulting combined key is at least as secure as the most robust of the two original keys. Therefore, an attacker would have to successfully decipher both the quantum key distribution and the public-key cryptography in order to intercept the combined key. The combined key is used in a high-capacity symmetric cipher to encrypt a wide frequency range network connection, and it changes frequently and quickly (Lee et al., 2006).

Although the security of using a shorter key in a symmetric cipher is not absolute, the main reason for adopting this architecture is the seamless integration it offers with existing networks. Customers have become familiar to the widespread use of classical cryptography, which is capable of efficiently encrypting large network links.

While it is true that using a key shorter than the encrypted message in a symmetric cipher does not provide unconditional security, the decision to adopt this architecture is primarily driven by the need for easy integration into existing networks. Customers are used to the availability of classical cryptography that can effectively encrypt the entire gigabit network link.

Commercially available devices have a key generation speed and average bandwidth for one-time pad that does not exceed a few kilobits per second. However, experimental prototypes have shown speeds of up to 1 megabit per second over a 50 kilometer fiber (Amir et al., 2004).

8.6.2. Quantum Key Distribution Networks

In order to expand the user base and address the constraint of connection distance, there are two potential network configurations: one using trustworthy nodes and the other involving untrusted nodes.

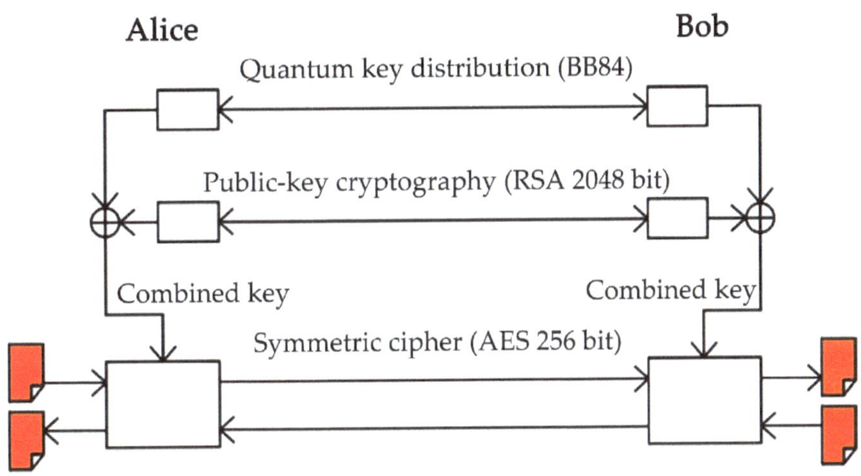

Figure 8.6. Schematic of commercial quantum cryptography vintage 2010.

Source: Dag Roar Hjelme, Creative Commons License.

The trusted-node network comprises point-to-point quantum key distribution links connecting the nodes. In order to create a shared key, two users determine a path through intermediate nodes. Then, using keys created at each point-to-point link along the way, one user uses a series of one-time-pad encryptions to transfer their key to the other user. A number of urban areas have had this network design (Cao et al., 2022).

An uninterrupted optical channel can be established between end users in the untrusted-node network by utilizing optical switches at the nodes. This is a feasible option given the current state of technology. However, it's important to note that the optical switches do not enhance transmission distance, limiting their application to geographically confined networks. An further strategy makes use of quantum repeaters at the untrusted nodes, which might greatly increase the distance beyond the existing 250 kilometer limit (Peev et al., 2009).

Nevertheless, quantum repeaters are still a technology that lies ahead in the future. The network configuration involving untrusted nodes has the potential to fully utilize the advantages of quantum cryptography, potentially surpassing the use of trusted couriers and other methods for key distribution. As an example, every user has the ability to gather and keep only the initial authentication keys for other users on the network. They can then generate additional key material with any user whenever necessary (Mehic et al., 2020).

ACTIVITY 8.1.

Alice and Bob have obtained a raw key of considerable length, consisting of 1,000 bits, following strict error correction procedures. It is estimated that a certain party possesses knowledge of approximately 10% of the key bits. What is the appropriate number of bits to retain after privacy amplification in order to guarantee a secure final key?

SUMMARY

- The potential of quantum cryptography in key distribution networks has been established through successful demonstrations at lengths of up to 250 kilometers. Despite current limitations in transmission rates, these systems offer a reliable method for secure communication in cases where the existing public-key systems may not be considered trustworthy. However, it is important to note that quantum cryptography is being developed with a focus on the potential future feasibility of breaking classical public-key ciphers. As an example, the potential of a quantum computer to decipher modern encryption methods is an interesting prospect. Quantum cryptography is an excellent example of the close connection between fundamental and applied research.

REVIEW QUESTIONS

1. What is the main difference between the security basis of quantum cryptography and classical cryptographic techniques?
2. Explain the concept of measurement in quantum mechanics and how it affects the state of a qubit.
3. What is the primary use of quantum cryptography in secure communication?
4. Describe the BB84 protocol for quantum key distribution.
5. Explain the processes of error correction and privacy amplification in the context of quantum cryptography.

MULTIPLE CHOICE QUESTIONS

1. A quantum bit (qubit) can be represented by:

a. An electron.

b. A proton.

c. A photon.

d. A neutron.

2. Which theorem states that qubits cannot be perfectly copied?

a. No-deletion theorem.

b. No-hiding theorem.

c. No-cloning theorem.

d. No-splitting theorem.

3. The BB84 protocol for quantum key distribution was developed by:

a. Stephen Wiesner and Charles Bennett.

b. Charles Bennett and Gilles Brassard.

c. Stephen Wiesner and Gilles Brassard.
d. Albert Einstein and Niels Bohr.

4. **In the BB84 protocol, which states are primarily used to encode the qubits?**
   a. Energy states.
   b. Position states.
   c. Spin states.
   d. Polarization states.
5. **In the context of quantum cryptography, what is an eavesdropper commonly referred to as?**
   a. Alice.
   b. Bob.
   c. Eve.
   d. Charlie.
6. **Which process ensures that the final key is error-free and secure after detecting imperfections during transmission?**
   a. Key swapping.
   b. Privacy amplification.
   c. Quantum entanglement.
   d. Error correction.

Answers to Multiple Choice Questions

1. (c); 2. (c); 3. (b); 4. (d); 5. (c) 6. (d)

REFERENCES

1. Amir, Y., Kim, Y., Nita-Rotaru, C., & Tsudik, G. (2004). On the performance of group key agreement protocols. ACM Transactions on Information and System Security (TISSEC), 7(3), 457–488.
2. Babinec, T. M., Hausmann, B. J., Khan, M., Zhang, Y., Maze, J. R., Hemmer, P. R., & Lončar, M. (2010). A diamond nanowire single-photon source. Nature Nanotechnology, 5(3), 195–199.
3. Bai, S., Lepoint, T., Roux-Langlois, A., Sakzad, A., Stehlé, D., & Steinfeld, R. (2018). Improved security proofs in lattice-based cryptography: Using the Rényi divergence rather than the statistical distance. Journal of Cryptology, 31, 610–640.
4. Barkadehi, M. H., Nilashi, M., Ibrahim, O., Fardi, A. Z., & Samad, S. (2018). Authentication systems: A literature review and classification. Telematics and Informatics, 35(5), 1491–1511.
5. Barnes, W. L., Björk, G., Gérard, J. M., Jonsson, P., Wasey, J. A. E., Worthing, P. T., & Zwiller, V. (2002). Solid-state single photon sources: Light collection strategies.

The European Physical Journal D-Atomic, Molecular, Optical and Plasma Physics, 18, 197–210.

6. Barnum, H., Nielsen, M. A., & Schumacher, B. (1998). Information transmission through a noisy quantum channel. Physical Review A, 57(6), 4153.
7. Basu, S., Fernald, J. G., & Kimball, M. S. (2006). Are technology improvements contractionary? American Economic Review, 96(5), 1418–1448.
8. Becker, K. H., Schoenbach, K. H., & Eden, J. G. (2006). Micro plasmas and applications. Journal of Physics D: Applied Physics, 39(3), R55.
9. Bennett, C. H., Brassard, G., & Ekert, A. K. (1992). Quantum cryptography. Scientific American, 267(4), 50–57.
10. Brassard, G., Chuang, I., Lloyd, S., & Monroe, C. (1998). Quantum computing. Proceedings of the National Academy of Sciences, 95(19), 11032–11033.
11. Broadbent, A., & Schaffner, C. (2016). Quantum cryptography beyond quantum key distribution. Designs, Codes and Cryptography, 78, 351–382.
12. Cao, Y., Romero, J., Olson, J. P., Degroote, M., Johnson, P. D., Kieferová, M., & Aspuru-Guzik, A. (2019). Quantum chemistry in the age of quantum computing. Chemical Reviews, 119(19), 10856–10915.
13. Cao, Y., Zhao, Y., Wang, Q., Zhang, J., Ng, S. X., & Hanzo, L. (2022). The evolution of quantum key distribution networks: On the road to the qinternet. IEEE Communications Surveys & Tutorials, 24(2), 839–894.
14. Choe, J. S., Ko, H., Choi, B. S., Kim, K. J., & Youn, C. J. (2018). Silica Planar Lightwave Circuit Based Integrated 1×4 Polarization Beam Splitter Module for Free-Space BB84 Quantum Key Distribution. IEEE Photonics Journal, 10(1), 1–8.
15. Chuang, I. L., & Yamamoto, Y. (1996). Quantum bit regeneration. Physical Review Letters, 76(22), 4281.
16. Collins, M. J., Xiong, C., Rey, I. H., Vo, T. D., He, J., Shahnia, S., & Eggleton, B. J. (2013). Integrated spatial multiplexing of heralded single-photon sources. Nature Communications, 4(1), 2582.
17. Czermann, M., Trócsányi, P., Kis, Z., Kovács, B., & Bacsárdi, L. (2021). Demonstrating BB84 quantum key distribution in the physical layer of an optical fiber-based system. Info Communications Journal, 13(3), 45–55.
18. Devetak, I. (2005). The private classical capacity and quantum capacity of a quantum channel. IEEE Transactions on Information Theory, 51(1), 44–55.
19. DiVincenzo, D. P., Shor, P. W., & Smolin, J. A. (1998). Quantum-channel capacity of very noisy channels. Physical Review A, 57(2), 830.
20. Duc, A., Faust, S., & Standaert, F. X. (2019). Making masking security proofs concrete (or how to evaluate the security of any leaking device), extended version. Journal of Cryptology, 32(4), 1263–1297.
21. Fischlin, R., & Schnorr, C. P. (2000). Stronger security proofs for RSA and Rabin bits. Journal of Cryptology, 13, 221–244.
22. Grassl, M., Rötteler, M., & Beth, T. (1998). Computing local invariants of quantum-bit systems. Physical Review A, 58(3), 1833.

23. Holevo, A. S. (1998). The capacity of the quantum channel with general signal states. IEEE Transactions on Information Theory, 44(1), 269–273.
24. Jain, N., Anisimova, E., Khan, I., Makarov, V., Marquardt, C., & Leuchs, G. (2014). Trojan-horse attacks threaten the security of practical quantum cryptography. New Journal of Physics, 16(12), 123030.
25. Jennewein, T., Achleitner, U., Weihs, G., Weinfurter, H., & Zeilinger, A. (2000). A fast and compact quantum random number generator. Review of Scientific Instruments, 71(4), 1675–1680.
26. Kak, S., Verma, P., & MacDonald, G. (2011). Cryptography and system state estimation using polarization states. In The Nature of Light: What are Photons? IV (Vol. 8121, pp. 225–233).
27. Khaleel, A. I., & Tawfeeq, S. K. (2009). Real time quantum bit error rate performance test for a quantum cryptography system based on BB84 protocol. Iraqi Journal of Laser, 8(A), 29–35.
28. Komiyama, S., Astafiev, O., Antonov, V., Kutsuwa, T., & Hirai, H. (2000). A single-photon detector in the far-infrared range. Nature, 403(6768), 405–407.
29. Kumar, P., Gurtov, A., Sain, M., Martin, A., & Ha, P. H. (2018). Lightweight authentication and key agreement for smart metering in smart energy networks. IEEE Transactions on Smart Grid, 10(4), 4349–4359.
30. Lee, C., Sohn, I., & Lee, W. (2022). Eavesdropping detection in BB84 quantum key distribution protocols. IEEE Transactions on Network and Service Management, 19(3), 2689–2701.
31. Lee, P. P., Lui, J. C., & Yau, D. K. (2006). Distributed collaborative key agreement and authentication protocols for dynamic peer groups. IEEE/ACM Transactions on Networking, 14(2), 263–276.
32. Lo, H. K., & Chau, H. F. (1998). Why quantum bit commitment and ideal quantum coin tossing are impossible. Physica D: Nonlinear Phenomena, 120(1–2), 177–187.
33. Lyu, S., Rockmore, D., & Farid, H. (2004). A digital technique for art authentication. Proceedings of the National Academy of Sciences, 101(49), 17006–17010.
34. Marsaglia, G., Zaman, A., & Tsang, W. W. (1990). Toward a universal random number generator. Statistics & Probability Letters, 9(1), 35–39.
35. Mason, W. P. (1981). Piezoelectricity, its history and applications. The Journal of the Acoustical Society of America, 70(6), 1561–1566.
36. Maurer, U., & Wolf, S. (2003). Secret-key agreement over unauthenticated public channels. II. Privacy amplification. IEEE Transactions on Information Theory, 49(4), 839–851.
37. Mehic, M., Niemiec, M., Rass, S., Ma, J., Peev, M., Aguado, A., & Voznak, M. (2020). Quantum key distribution: A networking perspective. ACM Computing Surveys (CSUR), 53(5), 1–41.
38. Molotkov, S. N. (2019). On the secrecy of a simple and effective implementation of BB84 quantum cryptography protocol. Laser Physics Letters, 16(7), 075203.

39. Murch, K. W., Weber, S. J., Macklin, C., & Siddiqi, I. (2013). Observing single quantum trajectories of a superconducting quantum bit. Nature, 502(7470), 211–214.
40. Najafi, F., Mower, J., Harris, N. C., Bellei, F., Dane, A., Lee, C., & Englund, D. (2015). On-chip detection of non-classical light by scalable integration of single-photon detectors. Nature Communications, 6(1), 5873.
41. Ofek, N., Petrenko, A., Heeres, R., Reinhold, P., Leghtas, Z., Vlastakis, B., & Schoelkopf, R. J. (2016). Extending the lifetime of a quantum bit with error correction in superconducting circuits. Nature, 536(7617), 441–445.
42. Peev, M., Pacher, C., Alléaume, R., Barreiro, C., Bouda, J., Boxleitner, W., & Zeilinger, A. (2009). The SECOQC quantum key distribution network in Vienna. New Journal of Physics, 11(7), 075001.
43. Pirandola, S., Andersen, U. L., Banchi, L., Berta, M., Bunandar, D., Colbeck, R., & Wallden, P. (2020). Advances in quantum cryptography. Advances in Optics and Photonics, 12(4), 1012–1236.
44. Pirandola, S., Laurenza, R., Lupo, C., & Pereira, J. L. (2019). Fundamental limits to quantum channel discrimination. npj Quantum Information, 5(1), 50.
45. Portmann, C., & Renner, R. (2022). Security in quantum cryptography. Reviews of Modern Physics, 94(2), 025008.
46. Ramos, M. F., Pinto, A. N., & Silva, N. A. (2022). Polarization based discrete variables quantum key distribution via conjugated homodyne detection. Scientific Reports, 12(1), 6135.
47. Reuss, R. H., Chalamala, B. R., Moussessian, A., Kane, M. G., Kumar, A., Zhang, D. C., & Snow, E. (2005). Macroelectronics: Perspectives on technology and applications. Proceedings of the IEEE, 93(7), 1239–1256.
48. Senellart, P., Solomon, G., & White, A. (2017). High-performance semiconductor quantum-dot single-photon sources. Nature Nanotechnology, 12(11), 1026–1039.
49. Shor, P. W., & Preskill, J. (2000). Simple proof of security of the BB84 quantum key distribution protocol. Physical Review Letters, 85(2), 441.
50. Singh, S. P. (2022). Quantum cryptography and its application. In International Conference on Computational Mathematics & Engineering Applications (ICCMEA) on June (Vol. 24, p. 26).
51. Somaschi, N., Giesz, V., De Santis, L., Loredo, J. C., Almeida, M. P., Hornecker, G., & Senellart, P. (2016). Near-optimal single-photon sources in the solid state. Nature Photonics, 10(5), 340–345.
52. Steane, A. (1998). Quantum computing. Reports on Progress in Physics, 61(2), 117.
53. Techateerawat, P. (2010). A review on quantum cryptography technology. International Transaction Journal of Engineering, Management & Applied Sciences & Technologies, 1, 35–41.
54. Tsurumaru, T. (2021). Equivalence of three classical algorithms with quantum side information: Privacy amplification, error correction, and data compression. IEEE Transactions on Information Theory, 68(2), 1016–1031.

55. Uppu, R., Pedersen, F. T., Wang, Y., Olesen, C. T., Papon, C., Zhou, X., & Lodahl, P. (2020). Scalable integrated single-photon source. Science Advances, 6(50), eabc8268.
56. Velásquez, I., Caro, A., & Rodríguez, A. (2018). Authentication schemes and methods: A systematic literature review. Information and Software Technology, 94, 30–37.
57. Wang, X. B. (2005). Beating the photon-number-splitting attack in practical quantum cryptography. Physical Review Letters, 94(23), 230503.
58. Watanabe, S., Matsumoto, R., & Uyematsu, T. (2006). Noise tolerance of the BB84 protocol with random privacy amplification. International Journal of Quantum Information, 4(06), 935–946.
59. Zappa, F., Lacaita, A. L., Cova, S. D., & Lovati, P. G. (1996). Solid-state single-photon detectors. Optical Engineering, 35(4), 938–945.
60. Zbinden, H., Bechmann-Pasquinucci, H., Gisin, N., & Ribordy, G. (1998). Quantum cryptography. Applied Physics B: Lasers & Optics, 67(6).

Index

FSC
www.fsc.org
MIX
Papier | Fördert gute Waldnutzung
FSC® C083411

Kolibri 360 GmbH
Ferdinand-Jühlke-Straße 7
99095 Erfurt, Deutschland
produktsicherheit@kolibri360.de